Due Diligence
Memoirs of the Life of
an Engineer and Outdoorsman
Book 1

Dwight Lee Bates PE

Copyright © 2014 Dwight Lee Bates

All rights reserved.

ISBN:10:1503029514

ISBN-13: 978-1503029514

Dwight Lee Bates

DEDICATION

I dedicate this book to all those who love Engineering and the Outdoors like I do. Every effort was made to be factual and honest in this book.

Introduction

I wrote these books to highlight my accomplishments as an Engineer. I felt since I won 207 awards and saved $26 million in my 44 year career I should document how I accomplished it through due diligence. My ex boss at Boeing at the Composite Manufacturing Center summed it up at my second retirement party. He said: "Lee wrote notes in a notebook which expanded as the project went on. He did not miss anything and wrote everything down. This is why he has won so many awards." The first book I wrote was entitled: "Lee's Life" in the standard memoirs format. The second was "Poetic Justice" on poems about my life that I wrote in Ode format. The third book for my memoirs is 20 stories of light reading about my Engineering accomplishments plus 113 light reading stories mostly on the Outdoors. The fourth, fifth, sixth, seventh and eighth memoirs books are various interesting topics and meant to be light reading. These books average about 130 stories in each book. My friends have pushed me into writing these memoirs since they said I have led such an interesting life. While working for Boeing, I went to Middle Schools to try to convince our children to become Engineers. Our country needs them. This way we do not have to hire foreign Engineers. I thought that writing these memoirs would be a good way to get to them. Also I went on hundreds of outings in the beautiful Northwest. I thought these memoirs would be a good way to get our children outdoors. They only have to read these memoirs to find out how to do what I did. I loved every minute. I wrote these memoirs at the 6th grade level to inspire these kids. My nephew Jon was inspired by the stories I told him on mountain climbing. Some of those I put into these memoirs. Jon wrote: "Listening to you, dad and Uncle Jay talk about climbing around the campfires growing up was my main motivation to want to climb as much as I have. I would read the books you would circulate to my dad and dream about the adventures to be had. I appreciate the childhood I had and all the time I spent in the outdoors. I hope I can do the same with my daughter. Thanks." Jon went on to lead climbers on Mount Hood in Oregon. He has climbed on El Capitan in Yosemite and Mount Stuart as I wrote in these memoirs. Maybe after I am committed to ashes, other people will do the same memorable things that I did by following my lead. I hope this is my legacy.

CONTENTS

1	Dedication	2
2	Introduction	2
3	SAILING	15
4	Live Aboard	16
5	Dusty	19
6	Hydos	22
7	Jet Ski	23
8	The Sailboat Race	25
9	SKIING	31
10	Sun Mountain	32
11	Snow Cave	35
12	Office Cross Country	37

Dwight Lee Bates

13	Mission Ridge	39
14	Slush Cup	41
15	Grizzly Slide	42
16	CAREER	46
17	The Jet Age	47
18	The Boeing Supersonic Transport	49
19	The Best Engineering I Ever Did	55
20	My Flight Instructor Dick Pingrey's 747 Engine Problems	60
21	The Boeing 747 Water Spray Test	61
22	Lee's Career as an Engineer at Boeing	65
23	MRB	68
24	Lee's 2013 Resume	71
25	Puget Sound Naval Shipyard	74

Dwight Lee Bates

26	The 50 Percent Review Conference	75
27	Marketing	77
28	The MiniWinch	78
29	Lee Saves Boeing Ten Million Dollars	80
30	The Resource Center for the Handicapped	81
31	The 1969 Boeing Incredibles	85
32	777 Design Build Team	88
33	Why I Want to Return to Boeing as an Engineer	91
34	Boeing Composite Manufacturing Center	94
35	Boeing 787 Wins Collier Trophy	95
36	Jet Liner Bail Out	96
37	Telemetry	98
38	Early Retirement	100

Dwight Lee Bates

39	Aussies	102
40	Presentations	105
41	CAD	106
42	FLYING	110
43	Young Eagles	111
44	A B-17 Saved My Life	114
45	Flying Scares	117
46	Tail Wheel Endorsement	120
47	My First Flight of My Airplane I Built	123
48	Boeing Model 40 and 787	126
49	Gyro Copter Flight	130
50	Slipstick	132

Dwight Lee Bates

51	C-46	13
		4
52	Jack	13
		6
53	Morgan Marry Me	140
54	<u>Airplane Mysteries</u>	143
55	D B Cooper	143
56	Amelia	143
57	Missing 777	147
58	Lake Cheakamus	150
59	F-18	153
60	Ploesti	155
61	Night Flying	157
62	Plane Crashes on Elk Mountain	159

Dwight Lee Bates

63	CARS	164
64	<u>1967 Cougar</u>	165
65	1967 Cougar on KIMA TV	165
66	Lee's 1967 Cougar Facts	165
67	Lee on 1967 Cougar on KOMO Radio	169
68	Lee on TV and Radio	171
69	Engine Swap	172
70	51 Chevy	175
71	<u>Racing</u>	177
72	Formula One	177
73	Senna	177
74	Indy	178

75	PERSONAL	183
76	Our Wedding	184
77	Our House	186
78	Our View	186
79	Our Yard	187
80	My Shop	187
81	MBA	194
82	The Professor	197
83	Dreams	200
84	Viet Nam	203
85	Aging	207
86	Les Miserable's	209

Dwight Lee Bates

Page	Title	Page
87	Why I Write Poetry	211
88	The Ellensburg Address	212
89	Grandpa's Poems	213
90	Ma's Lineage	221
91	Bungee Jumping	226
92	Dad's 80th	228
93	Tomahawk	231
94	Diane's College Degree	233
95	THE OUTDOORS	235
96	Lee's Search and Rescue Help	236
97	Solo Climbing Three Fingers Mountain	237
98	Lee's Jump	239

Dwight Lee Bates

99	Kayaks and Sailing Dinghies	243
100	Rampart Ridge	250
101	Long Draw	252
103	Rainbow and Therriault Lakes	254
103	Lake Hamby	257
104	Climbing the Brothers	259
105	Mildred Lakes	270
106	Link Lake	274
107	Cutthroat	276
108	Mulie	281
109	Sisters	283
110	Baca's Boat	288

Dwight Lee Bates

111	Dewey Lake	290
112	Owyhee River	295
113	Mount Stuart	298
114	Wild Fire Fighting	303
115	<u>California</u>	311
116	Shasta	311
117	Sierras	311
118	Fred and Dorothy's Cabins	316
119	Ingalls Lake	318
120	Tamarack	323
121	Lake Melakwa	325
122	ROKHOUNDING	327

Dwight Lee Bates

#	Title	Page
12	Rockhounding	323
123	Picture Agate	330
124	Hampton Butte	332
125	NEAT PLACES	334
126	Elk Mountain	335
127	Fort Halleck	339
128	Bloody Lake Massacre	342
129	The Smokejumper Reunion	346
130	Elk Camp	348
131	Great Falls	350
132	Northwest Trek	353
133		

ACKNOWLEDGMENTS

I acknowledge the help of my brothers Jay and Scott who reviewed this book and gave valuable comments. Also Wikipedia for clarification of my stories.

Dwight Lee Bates

SAILING

Live Aboard

From 1974 to 1978, my wife, our cat Pumpkin and I lived aboard our new sailboat the Ranger 29 called Catnip shown in the following photo. For what it looked inside the sailboat see another story in these memoirs called: " Our Sailboats and Cruises." We took showers in the marina restrooms ashore. It was a brand new marina run by the Port of Port Orchard. The Harbor Master was Jim Armstrong who always appreciated our watching the marina at night. There was about 4 live aboard boats in the marina. They limited the number. We were the first to move aboard. Our sailboat was also brand new. It was nice to have a new sailboat, new marina and new job. I could see the marina from my office window where I worked in the Engineering Building in Puget Sound Naval Shipyard across Sinclair Inlet. I did Engineering on all the types of U S Navy ships. I commuted to work on the foot ferry. This was a 40 foot boat which went back and forth every day between Bremerton and Port Orchard. Brownie was the skipper and he used to let me drive the foot ferry. We were the official guards for the marina at night. If something was out of line, I could pick up the phone on the dock next to our boat. This went directly to the police station. I only had to do this phone call twice. Once when I thought someone was stealing another person's dinghy. Another time when I reported a boat on fire. I used the dock hose to put out the fire after I used the phone. Only later did the boat owner tell me he had a 50 gallon propane tank aboard. If this had exploded, ½ the marina would have blown up and probably I would have been killed. As it was, I poured water inside the boat until it almost sank. The sunsets from the back of our sailboat as shown in the other story were beautiful. We often cooked on the dock and ate in the sailboat cockpit with people we invited over for dinner. People were always giving us an extra Salmon they caught to cook for dinner. I never got a cold while living aboard. Maybe it was the fresh salt air. Everyone was in a good frame of mind when they came down to their boat at the marina. They were in a stress relieving mode. I would walk down the dock socializing with these people. Dusty lived not too far from us. He is another story in these memoirs. I loved living aboard. We also raced our sailboat as described in another story in these memoirs. I would move back aboard in a heart beat. I loved it so. It was our passion. It is hard to explain to others unless they have done it. I guess that is what I am trying to do with these memoirs. Also people do not know what Engineers do and how they think. My publisher liked my Engineering poems to publish.

Maybe this is because Engineers are such a mystery. We do good things but never brag about them. I do not think writing my memoirs is bragging but it may sound that way. These memoirs are a source of unlocking what Engineers do and how they think. We are a different breed of cat. This is shown by our living aboard. People used to say; " I do not see how you can live in such a cramped space?" The answer is our table folded up against the bulkhead to give us room. And we cruised almost every weekend to another place. We did not spend that much time aboard the sailboat. It was great sleeping and the waves would rock you to sleep.

Dwight Lee Bates

Dusty

A fellow live aboard in the Port Orchard Marina was Dusty. He lived on an old boat without a mast that you could not stand up in. He had a wood stove to keep warm. He said that he had been in trouble with the law. He helped me paint my 1967 Cougar which is what he did for a living. He used to want to go out when it was blowing whitecaps. We would take his boat out in storms and loved it. People said it was too dangerous. He had a little 5 year old boy who used to visit him. We would feed him humbergers as he called them. We used to spend many a night sitting in the cockpit of our boat watching the sunset as shown in the following photo I took. The other following photo was taken of our live aboard sailboat Catnip, me and our dinghy Catnap from the top of a mast when a friend went up it to do some repairs. After we left the marina, we heard from other people that Dusty motored down the coast to hide from paying off the bank for his boat. But he got caught they said.

Dwight Lee Bates

Dwight Lee Bates

Hydros

I remember going to see the hydros on Lake Washington. They were thrilling because of the sound of the World War 2 airplane engines. The hydros then were made out of wood. Now a days the big roar of the hydros is replaced by the whine of a helicopter jet engine. It is not the same. I remember 2 hydros colliding at the Sea Fair Race right in front of us. You saw a lot of spray and then floating pieces of wood. Then you could just make out 2 heads bobbing in the water. No one was killed. In those days, they were not strapped in but now a days it is different. The cockpit is a capsule that separates from the rest of the boat in the event of a crash. This capsule even has an oxygen supply for the driver. The modern hydros are made of fiberglass and not wood. The danger now a days is flipping over. The hydro rides on the prop and the 2 sponsons. If you get too much air under it, up you go into the air. The results of colliding with the water at those speeds is not good. Although some boats have gone airborne and landed back on the water without damage, it does not happen often. I like it when between the races they run the old hydros. I remember the old Slo Mo Shun on Lake Washington in about 1956. It was black and white TV then. I have watched the hydros every year since then. My favorite driver was Bill Muncey who got killed in a race.

Jet Ski

In 1995, I rented a jet ski at Chelan on Lake Chelan. The first one I rented you had to stay in the water so you could ride it. It drug you along in the water. So I took it back and rented the one shown in the following photo. I found nothing exciting about a jet ski. I would rather be water skiing where it takes skill. The jet ski to me is either for those without skill or are afraid to water ski. I ran around in circles and tried to jump my own wake. However I took it back before my time was up. It was just too boring. I could see someone buying a house in a cove for rest and relaxation only to have some young kid roar around on a jet ski all day. I have had the same trouble with young kids roaring around on ATVs all day. I was there for peaceful camping. This happened a week ago on the Owyhee River when kids were going 50 miles per hour on noisy ATVs. If they crashed, who was going to help them miles from their camp? I think parents give their kids ATVs just to occupy them. The only good use for a ATV in my mind is for a farmer or rancher to use them for small chores. And I helped design the small electric winches on the front of them. I see a snowmobile as useful to a hunter and trapper but I was bored with one at Dixie. It was a good way to get around in the backcountry. I felt the people we visited in Dixie, Idaho on snowmobiles resented our noise. What every happened to peace and quiet in the woods like you get on cross country skis?

Dwight Lee Bates

The Sailboat Race

Our regatta was 3 races on a Saturday at Sinclair Inlet, Port Orchard. It was sponsored by the local Yacht Club - the Port Orchard Yacht Club. I represented the Shilshole Bay Yacht Club from Seattle. I had a crew of 5 on our Ranger 29 foot sailboat. It had a inboard engine in it. I would check the weather report for Saturday morning. My crew would show up and we would motor out to the start line. I rigged up the sailboat before hand. Terry and Dick worked with me at the shipyard. We would sail around until we heard the 5 minute gun. Then I would take up the windward side of the start line. This would allow me the cleanest air. I would let the sails luff as I would block this best start position. At 10 seconds to go, I would pull in the sheets to harden up the sails. When the start gun went off, I had the best position. I won 13 straight starts this way until other people tried it too. Going upwind, you could have walked from deck to deck since there was only 10 feet between each of the 7 sailboats. When my brother in law Arnie crewed for me, he was amazed when he saw this. He said I got him started in sailing when I kept my 11 foot sailboat at his house on Lake Washington. He bought 3 sailboats to our 5. At the windward mark, I would tell my foredeck man to attach the spinnaker bag to the pulpit. Then as we rounded the mark, the foredeck man would set the spinnaker pole. Then we would raise the spinnaker outside of the head sail. The sheets for both sails had to be routed just right. As the spinnaker would fill, I would jibe the main sail boom to run down wind. Going down wind, you would watch the tell tale at the top of the mast. You would keep the spinnaker pole and main sail boom perpendicular to it. You would look for puffs and jibe to get the best wind. You had to watch out for other boats and know the right of way rules. At the leeward mark, you would reverse all the steps and raise the head sail inside the spinnaker. Our spinnaker had a cat on it that I sewed into the sail. Going upwind, you would look for dark water to sail into to get a lift. You would line up the tell tales on each side of the head sail to get full power. I tried to get the sailboat to heel at exactly 23 degrees to get the maximum speed. I would watch the knot meter, inclinometer, tell tales and depth sounder. I had to tell my crew what to do without seeming like Captain Bligh. We used to practice week nights to hone our skills. We would put our spinnaker up near the ferry so Diane could see it returning from work. My crew loved the sailboat and we were the talk of the office. I was called the "Sailor" at work. In the regatta, you did this for 3 races. My crew was exhausted at the end of the regatta.

During one regatta, the butter dish fell out of a cabinet and broke. I got a lecture from Diane over that one. One sailor who I worked at the shipyard sailed and raced his T Bird. He would visit the marina and tell my wife Diane what I did wrong during the regatta. She did not like this. He would even comment on my sailing my 8 foot dinghy out in the inlet in rough weather. He tried to get me to hit the committee boat during a race by not yielding right of way. He would collide with other boats on purpose to win a protest and a race. I thought it was great and used to call him to talk strategy. He knew a lot about racing and I had helped build a wooden T Bird sailboat. Before one regatta, I called the weather report and it was 60 mile per hour gusts. My crew and I chickened out and retired to the local pub located on the shore of the inlet. Here we drank beer and looked out the window at the regatta racing as a huge gust knocked down every sailboat. When they came back in, you should have seen the disgust on their faces as sails were torn and a lot of equipment was damaged. They said we were smart to stay ashore. The regatta was cancelled since no sailboats remained without damage. Usually 20 sailboats raced. People watched the races from shore including those from my office. On the exciting over night races on Puget Sound, we would rotate the crew. Three people were topside as the rest slept. We had a kerosene heater for warmth. We would sail in close to the shore to get out of the tide when it was against us. You kept your running lights on so other boats, ferries and ships could see you. Racing in the fog was an adventure. I loved every minute. As Ted Turner said after winning the Americas Cup: " If you can race a sailboat, you can do anything."

Dwight Lee Bates

Dwight Lee Bates

SKIING

Sun Mountain

In 1998, I went to Sun Mountain Lodge near Winthrop to cross country ski with my friend Mike Harper and his son Justin. We stayed in a motel in Winthrop right on the Methow River which I had river rafted before. This is another story. We got up early, packed a lunch, ate a big breakfast and headed for the slopes. The Sun Mountain Lodge as shown in the following photo is huge and expensive. We toured through the lodge but after looking at the expensive menu, we decided we could get more bang for the buck in Winthrop. Sun Mountain must be for doctors, lawyers and dentists from Seattle and not lowly Boeing Engineers like us. Our cheaper motel room did just fine. We cross country skied to the top of the pass. The view up there was tremendous as shown in the following photo. The run to the lodge was 3 miles and all downhill. I was going 30 miles an hour in the steep areas. I skied it nonstop in about 20 minutes. Mike had to keep track of Justin so we then skied separately. Mike and Justin also went cross country skiing with me at Scottish Lakes where we went in by snow cat. That is another story in my memoirs. That night we went out on the town for a big dinner. Then we adjourned to the motel hot tub. There is nothing like sitting in a hot tub and telling wild stories after a hard days skiing. The hot tub at Scottish Lakes floated in the water with you. To find out, read another story in my memoirs. We then drove home in Mike's pickup. We had a great time and it did not cost us much.

Dwight Lee Bates

Dwight Lee Bates

Snow Cave

In 1996, I took my niece Emily and her boy friend to Paradise on Mount Rainier to build a snow cave. We put on snow shoes and slogged up to where Boy Scouts were building igloos. They looked to be much harder to build but had more room. I shoveled in a snow bank for 2 hours until I had a snow cave big enough for the 3 of us. When I came out, Emily's boy friend was no where to be found. Then I heard someone yelling a mile away on the ridge above us. It was her boy friend who yelled to Emily to come up there. We tried the snow cave. It was warm but the water kept dripping down on us as shown in the following photo. Then I noticed that Emily just had on a light jacket. She would have froze. So we abandoned our snow cave and snow shoed around to get warm. People have built snow caves to survive in when they are hit by a blizzard. It was about 40 degrees inside. You would not freeze but you probably would not be too comfortable. But if it saved your life, it would be no problem. The next time I will build an igloo. Maybe I will earn a Boy Scout Badge. I built one in my yard and it was a lot of fun.

Dwight Lee Bates

Office Cross Country

In 1991, I invited my office to Ski Acres for cross country skiing. This was the third time I had done this. I take the entire group of 9 out to the practice area. I show them how to stop and turn. Many have never skied before. The final test is going down a slope. If you make it to the bottom without falling down, you pass. One girl in the group crashed. The group of 9 is shown in the following photo. As shown in the photo, my brother Scott went with us since he was in Puyallup on a Training Class for the Forest Service. Fun was had by all and no body got hurt. We usually adjourn to a friend's cabin for hot spiced wine but he was not with us this time. I told people there I was going to leave them and go to a new job. I would be the Boeing Liaison Representative for designing and building the 777 airplane in Renton. It was a sad farewell but I had to move on to bigger and better things. I have not done this since and I am now retired so maybe it will not happen again. It was fun while it lasted.

Dwight Lee Bates

Mission Ridge

In 1968, Diane and her office partner Jerry and their friends and I went skiing at the Mission Ridge Ski Area. We all went to the top of the area where the following photo was taken. It was icy cold. The hoar frost coated everything. We skied down to the B-24 Bomber that crashed in the ski area in World War 2. Only the wing is left. This ski run is called Bomber Run. We stayed in a motel in Wenatchee and went out to dinner. We can see the top of the run from our view in Ellensburg. I have flown over the ski area and looked down at the skiers. Also I saw where snowmobiles had contests to see how high they could get up on the ridge before they stopped. The Mission Ridge Ski Area is a real bargain on price but it is cold. The sun is always behind the ridge unless you are on top. From the area where the photo was taken, you can see the town of Wenatchee and way up the Columbia River. We have not seen Jerry and Larry for a long time and have moved on to other things. But we remember the ski area since we see it every day out our view.

Dwight Lee Bates

Slush Cup

In July 1972, we went to the Slush Cup at the Mount Baker Ski Area. You hike for about a half mile to a pond near a hill covered in snow. Their object was to drink a lot to gain courage. Then you ski down the hill fast enough so that your snow skis act like water skis and you make it across the freezing pond. Most people get ½ way and fall down in the freezing water. Some people ski down naked. It looked like to me all they wanted to do was get drunk. The winner got ¾ of the way across the pond, did a flop and landed back on his skis. He hit the other side and both skis came off. He landed flat on his back unhurt and had made it all the way across. This event has been copied in Alaska as I recently saw on TV. Warren Miller, the ski movie king, said the following about the Slush Cup: "The Mount Baker 2012 Slush Cup continues the tradition sparked up in the early1950's. On July 4th of 1953, I went to Mount Baker, Washington to film the Slush Cup. This was in the days when it was a real glacial pond and before it got shut down because of too many people trying for greatness while using chemicals in place of courage."

Grizzly Slide

The Grizzlies do not know what mountain climbers who glissade down a mountain know. Before you glissade, make sure you are not above a steep cliff with a drop off that will kill you. Sometimes you get going so fast that you can not stop. Remember the guy who built the cabin and lived in Alaska on his own. It was called: "Alone in the Wilderness." He filmed a Wolverine sliding down a snow slope for fun. Humans use snow boards so why not Grizzlies using their butt? The following photo shows a pursuing Mountain Lion and Mountain Sheep that slid down a snow slide and were killed when they hit the highway. The source of the photo is montanaoutdoor.com in 2014.

Lee

On 5/24/2014 9:33 PM, Scott Bates wrote:

Check out the pictures on my Facebook page of our hike we did in Glacier Park yesterday. We saw a sow and her one year old cub slide down a snow field like kids on a sled. They looked like they were having fun. We weren't quick enough to catch them on video. I read in the paper this morning that the Park Service snow plowing road crew found a dead one year old grizzly that had slid to its death in the same area on the Going To The Sun Highway a couple of days ago. The Fish and Wildlife had to do an autopsy on it since it is still a threatened species. They concluded that it had died from a fall off of a cliff below a snowfield because of the injuries to its head and ribs rather than a fight with another grizzly. Evidently, grizzly claws aren't nearly as good as ice axes but their wide rumps make good sleds. The grizzlies we saw were all feeding on the vegetation in avalanche chutes just below the snow line on the slopes. We saw Harlequin Ducks in McDonald Creek which is one of the few breeding areas for these beautiful ducks. We saw some goats too. Nice day, beautiful hike with a lot of wildlife sightings. All species including man were enjoying the spring weather. Spring is a good time to see lots of wildlife in the park. Scott

Dwight Lee Bates

Dwight Lee Bates

CAREER

The Jet Age

As I research Fort Halleck and the history of Elk Mountain, I imagine a couple riding in the seat of a wagon on the Overland Trail in 1864 like the following photo from Wikipedia. They are headed to California to start a new life. The husband says to his wife: " In 100 years a person will be born and raised in this land that we are traveling on. He will study to be an Engineer and help build 10,000 airplanes. These airplanes will cover the 2,000 miles of the trail we are traveling in 3 hours. These airplanes will travel at 7 miles high to every major city in the world carrying people of all different races." The wife says: " You must have had too much whiskey in Whiskey Gap."

Dwight Lee Bates

The Boeing Supersonic Transport

In January 1966, I started out to Boeing fresh out of college to take a position at Boeing in Seattle on the Supersonic Transport in the Propulsion Staff Group. My 1951 Chevy threw a rod in Kemmerer, Wyoming in the middle of a snow storm. This is another story in my Memoirs called 1951 Chevy. A semi gave me ride into Kemmerer where I caught a train to Seattle. Our train hit a car in Boise so I went back to look at it. That was a mistake. I made friends with some people also heading to Seattle. We played cards the whole way. I walked to a hotel in downtown Seattle from the train station. The next morning I caught a bus to Boeing to sign up for my job. The long line had 100 people in it . A Boeing Human Resources guy came by and said: " Any Engineers in line? " I said I was one and he said go to the front of the line. I signed up and borrowed $300 from payroll advance since I was broke. I found a place to live near the Boeing Plant 2 where the SST Propulsion Staff was located. I got a cheap studio apartment so I could pay off my student loans. I could lay in bed, watch TV and reach into the refrigerator in the tiny apartment. The bed made into couch. The only problem was when it was in couch mode it wanted to make back into a bed and when in the bed mode it wanted to make back into a couch. I borrowed $500 from my brother Scott to buy a 1964 Chevy Chevelle convertible. I loved that car. I was invited to dinner at woman's house I worked with and she introduced me to her beautiful young daughter. Obviously she thought I would make a good husband for her daughter. I had the same experience in Colorado so I ignored the daughter's advances. Also I had a blind date with a girl who worked nearby at Boeing. I did not hit it off with her. But she kept coming to my place of work to hustle me. My boss told me to knock it off and get to work. At my work, I was to locate the 4 big SST engines under the wing by locating the optimum pressure field by running a 1/16 scale SST model in the Boeing Supersonic Wind Tunnel. I used a flow field rake with total and static pressure probes, static ports and the Bernoulli Equation to locate the engines at 2.69 Mach Number. I also measured the boundary layer under the wing with a boundary layer rake I optimized from 3 different types of probes I designed. The purpose was to keep the engine flow out of the boundary layer or disturbed flow. After I ran the test, I had to reduce the data and write a report on my test results. I had to present my report to the Chief Engineer. He knew what I was doing and I was very impressed with his questions. He told me my test was very important. He was one smart guy. I remember the Supersonic Flow going through the wind tunnel test section hurt your chest and ears in the 5 seconds it blew down as we called it. Yes, I wore ear plugs. I took some Schlieren photos to verify that the shock waves stayed attached to the inlets. This was a light system shining through the test section where the denser air in the shock wave caused a shadow which we photographed as shown in the following photo. It took 3 hours to pressurize the 3 tanks that collected the pumped air for the blow down. You can see these tanks today from Highway 99 at Boeing Plant 2. I programmed the 1/16 SST model in a 3 Dimensional Potential Flow computer program to check my wind tunnel data. This program, I helped the programmer write, was loaded with key punched IBM cards and ran over night. If there a slightest error, the program would not run. This program was pretty accurate and showed how good my wind tunnel data was. I worked with my friend Phil Condit since he represented Aerodynamics. Later he became Boeing CEO and President. I was all for the SST variable sweep wing but we could not build a light enough wing pivot for the variable sweep wing. Also we could not machine Titanium efficiently. We used data from the secret B-70 and SR-71 Programs. I studied Secret engine inlet designs and we chose the translating spike conical inlet for the SST. This was similar to the SR-71 inlet. Then upper management decided against sweeping the wings. We had won the design competition from Lockheed because we swept the wings. Also my tests in the Boeing Transonic Wind Tunnel proved we had to sweep the wings. Then I heard the Boeing 747 Program was looking for Engineers, so I transferred to that program in the Propulsion Staff group. This was a good choice because the U S Government cancelled the SST Program after I left it. Also the 747 made history and I worked on it over a span of 30 years. On the 747 program, we were called the Incredibles because we got the job done in record time due to our hard work. In fact, I was honored recently by Boeing as an Incredible in 2009 at the 40th Anniversary of the 747 at a Boeing Museum of Flight Ceremony. The story of the Incredibles is in this book of my memoirs.

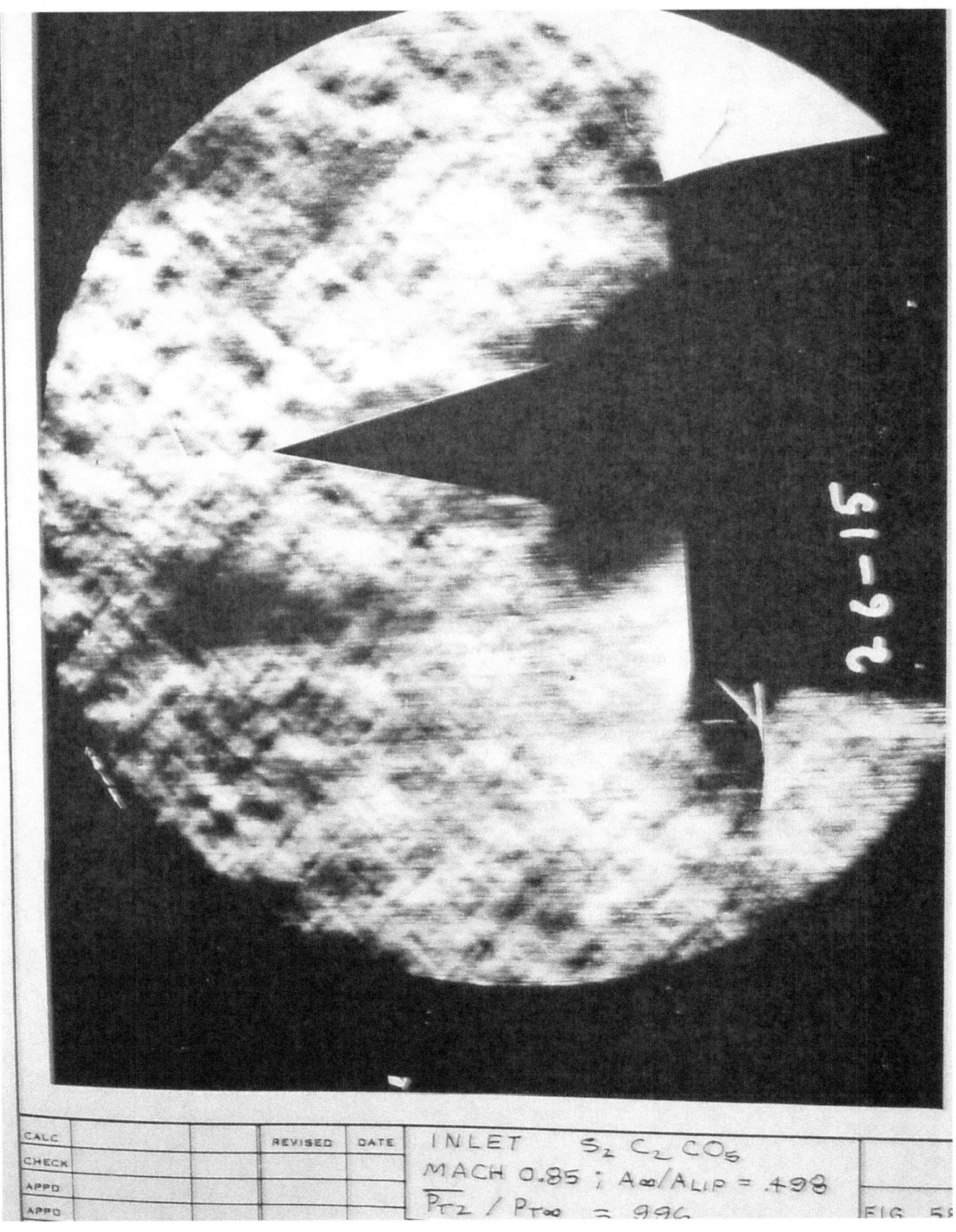

Dwight Lee Bates

```
'4767774767774767774767777477
"Wide Body Capital of the World"
         6-20-97
```

Phil,
I thought you might like to have this. I photographed it at the Painefield Boeing Restoration Center where it is hung on the wall.

Lee Bates
466 8892
RCH SST guy

Working Together
BOEING

June 27, 1997

Dear Lee,

Thanks for the great photo of the SST! Wow, that brings back memories!

Someday I'll have to visit the Restoration Center up there. Keep up the good work, and thanks for your contribution to the SST project as well as the 777.

I appreciate your thoughtfulness.

Dwight L. Bates
Mail Stop 0X-MX

(Phil Condit President and CEO the Boeing Company in 1997)

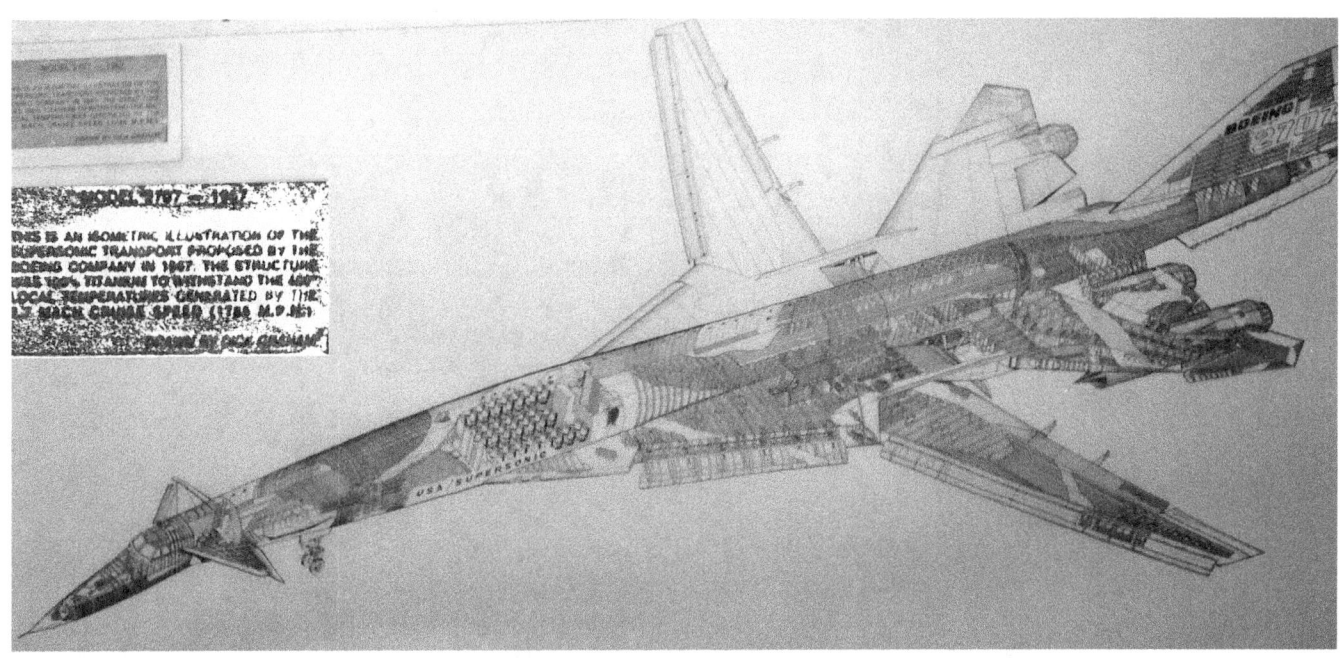

Dwight Lee Bates

The Best Engineering I Ever Did

In 1966, I had a good job working on the engines on Boeing's Supersonic Transport. It was why I was hired out of college in 1965. I enjoyed testing models of the SST in the Supersonic and Transonic Wind Tunnels at Plant 2. However due to problems I mentioned earlier, I decided to transfer to the Boeing 747 Program. As a result, I asked for and got an important job as a Propulsion Engineer on the 747 engines. The Boeing Company had bet the farm to build the 747 and the banks virtually owned Boeing. The 747 had to fly and sell or the Boeing Company would go bankrupt. I could not believe the responsibility that they gave me for flight testing the 747 engines. This story is on the problems we had certifying the 747 with bad engines. We could only test an engine for 20 hours on the average before we had to replace it. The entire future of the Boeing Company depended on certifying the 747 with bad engines. The following is the list of problems and how we overcame them. I verified my memory with information from two books on the 747: "Boeing 747, Design and Development Since 1969" by Guy Norris and Mark Wagner. The other book was the Boeing Chief Engineer's who is my friend and ex boss - Joe Sutter. His book is entitled: "747 Creating the World's First Jumbo Jet and Other Adventures from a Life in Aviation."

Inlet Blow In doors Causing Buzz Saw Noise

The engine was designed to run for most of its life at high cruise speeds and was therefore optimized by our short thin inlet. We ran many hours of wind tunnel tests to optimize the inlet. This created a problem at takeoff when a blunt edged inlet is better at allowing air into the engine at low speeds. To solve the problem, we adopted a set of 12 inward opening blow in doors around the mouth of the inlet. When full power was needed for takeoff, these doors opened to allow more air into the engine. These doors were closed at cruise when 500 pound springs closed the doors. This slender inlet was attached by a single point on top of the fan case which also caused a problem to be discussed later. Upon testing, we found the blow in doors caused a noise problem when the air flow separated off the doors and caused turbulent flow to hit the fan blades. This caused a noisy buzz saw noise similar to the noise of a table saw.

Solution - We redesigned a new inlet with a fatter lip to reduce the separation and eliminated the blow in doors. We optimized the inlet shape using a two dimensional potential flow program developed by Boeing.

Overheating Engines

The weight growth of the 747 caused the thrust requirement to balloon to 43,500 pounds of thrust when the original thrust requirement was 41,000 pounds. The increased thrust caused the turbine temperature limits to be exceeded. We went through 55 engines because we had to change them. The flight engineer's main job was to watch the engine temperatures. Pratt and Whitney who built the engines did not believe we had a problem until we took one of their top executives to 40,000 feet and had to shut down 3 of the 4 engines due to overheating. The engines would only last 20 hours on the average. Part of the problem was an insensitive barometric fuel control system. Normally the unit controlled fuel flow to the burners according to the pitot pressure at the inlet face, thus maintaining the correct fuel air ratio. As the system misread the conditions, it produced an excessively rich mixture which caused turbine overheating problems.

Solution - We adjusted sensitivity of the barometric fuel control system which made the engines run cooler and not exceed the limits.

Engine Surges

A big problem was due to the two concentric spools rotating at different rates could trigger a disruption of flow. This caused the fuel to bunch up and ignite in flame shooting out the front of the engine. It was similar to a car backfiring through the carburetor. It was spectacular at night and shook the whole airplane. The FAA Engineer was on board next to me when I surged the engine by going to max thrust reverse at 130 knots on landing. He said: " We realize you have a problem." The surge problem was also caused by ovalization of the engine case due to the way it was mounted. This caused the engine blades to rub on the side of the case and the engine to surge.

Solution - We fixed this by adding a strong back or Y shaped frame to support the engine case. Also Joe Sutter explains in his book that : "Changes to the engine surge valves and throttle control monitoring alleviated it sufficiently."

Five Percent Rise in Specific Fuel Consumption

The blades rubbing on the case problem caused a 5 percent increase in specific fuel consumption which was unacceptable to the airlines.

Solution - The strong back fixed this problem.

The Engines Would Not start

An electric motor turned the high pressure spool until light up speed was reached. However a tailwind caused the low pressure spool to rotate in the opposite direction to the high pressure spool. As a result, the airflow was wrong and the engine would not start.

Solution - Until Pratt fixed this, we parked the 747s facing into the wind.

The Lack of Thrust

We needed 45,000 pounds of thrust and only got 43,500 pounds.

Solution - We fixed this problem temporarily by water injecting the engine so we could get 45,000 pounds of thrust. Later the GE and Rolls engines had the right amount of thrust.

Cracked Engine Shafts

The reason we found out that we had a problem with cracked engine shafts was an engine exploded on the fight line in a run up test.

Solution - Pratt already knew about the problem and started sending us stronger shafts. We had to restack all the old engines and hung concrete blocks on the engine pylons until we could replace the engines. The Boeing Everett Plant looked like a glider factory.

Flutter Problem

During flight test at .86 Mach, the wing would flutter. When we first saw it, I realized that we had a problem which we reported.

Solution - We stiffened the wing with a redesign to fix the problem.

Paris Air Show

We wanted to show the 747 at the Paris Air Show in France but were not sure that the engines would last long enough to make it. I remember the decision came down to the Chief Test Pilot Jack Waddell who said in a meeting: " Lets see if we can make it to New York. If we make it that far with no problems, lets keep going."

Solution - They made the risky flight to Paris. The 747 was a hit at the Paris Air Show and the orders poured in.

Too Much Jet Blast

As we were taxiing out of the parking stall for a test, the pilot gunned the engines. I looked out the window and cars, people, line shacks and equipment were blown every where. I called on the intercom and said what had happened. When we landed after a 3 hour test and opened the door, George Nible the Everett Vice President was there. He told the pilot to apologize to the mechanics he blew around. Later a person was killed when they walked behind another 747.

Solution - Pilots were trained not to gun the engines taxiing out of the parking areas.

A 747 Hit's the Dike

A 747 hit the dike at Renton on Lake Washington when it came it for a landing with our Chief Engineer, Joe Sutter, on board. This was the same pilot that blew people around on our airplane.

Solution - The pilot was no longer with us after this.

Fireman's Pole

We were trained to bailout out if something went wrong by sliding down a fireman's pole to the cargo bay where the cargo door with a deflector would be opened. We walked around with parachutes at the ready. I loved sliding down the fireman's pole.

Solution - We never had to bailout but heard stories about an airplane getting upside down. The pilot told people to bail out but G forces held them in. The pilot recovered the airplane.

Stuck Engine Variable Blades

The variable engine blades would get stuck in the wrong position causing the engine to surge.

Solution - We got Pratt to fix this problem.

747 Model Airplane Joke

We heard arguing at a Boeing and Pratt management meeting so we put my 747 airplane model in through the door way. The arguing stopped as my model airplane rolled down the floor throwing sparks out the engines. Some one said look at the engines surging. This brought laughter and much relief. I still did not know if the Chief Engineer would fire me over my stunt so I never retrieved my model airplane. Finally he called me in and said Lee did you do that stunt? I said yes and he gave me my model 747 plane back. He said that was a good thing I did because it broke the tension and people got along better.

Solution - I never did that again.

Summary

The 747 was Certified on December 1969 by the FAA in part due to the Certification of the Engines Document I mainly wrote. I worked many hours overtime until midnight and conducted 40 hours of flight testing. It was the best Engineering I ever did in my career. We were called the Boeing Incredibles.

My Flight Instructor Dick Pingrey's 747 Engine Problems

As an addendum to the "Best Engineering I Ever Did" story in my memoirs, this story is about the guy who taught me to fly - Dick Pingrey. He flew the 747s for Pan Am with the bad Pratt and Whitney engines after the 747 was FAA Certified in 1969. He said that the flight engineer had to keep his hands on the four throttles and advance them slowly to prevent overheating the engines. He said Pan Am always had a 747 in reserve in case the they had to switch airplanes if the engines overheated. He said he was a flight engineer who had to constantly monitor the engine temperatures to keep them from overheating. He said it took awhile to fix all the problems with the engines. It is a small world that the guy, who would teach me to fly thirty years after I helped certify the engines, had to deal with the 747 Pratt and Whitney engine problems. I remember the first revenue flight across the Atlantic Ocean with the first 747. They loaded the 747 up with dignitaries and taxied out. Then the engines overheated so bad that they taxied back to the terminal. Then they unloaded the passengers and put them aboard another airplane. How embarrassing the issue was. I remember the long meetings with the engine manufacturer trying to straighten out the engine problems. Dick Pingrey later flew the 747 that I did the FAA Certification flight tests on. It was called the Clipper Storm since it was named after clipper sailing ships of the early 1800s. It was the second 747 off the Boeing production line. I flew in it 40 hours certifying the engines. Dick Pingrey was a flight engineer and pilot for Pan Am. It was run by Juan Tripp who kicked off the Boeing 747 program with his first order of 747s for Pan American World Airways.

The Boeing 747 Airplane Water Spray Test

In 1969, I was put in charge of the Boeing 747 Water Spray Test when I was a Senior Engineer at Boeing on the first 747 Program. During this test, we ran the 747 through 1 inch and 2 inch deep water in a trough on Boeing Field in Seattle. The test was to determine if the nose gear caused rain water to spray in a pattern into the engines which could lead to a loss of takeoff power. We coated the entire airplane with pink dye to determine where the water spray hit. We also filmed it from head on and from a camera on top of the fin. Our tests proved it was not a problem on the 747 which it was on the 727. On the Boeing 727, they had to install chinned tires on the nose gear to deflect the spray away from the engines. It was a fun test and a huge crowd would gather to watch it. This safety test was a part of the FAA Certification of the 747. I calculated the 747 would hydroplane at 50 miles per hour in the water trough. A friend of mine ran a flight school on Boeing Field. He saw the water spray test and told me about it. He was impressed. I told him that it was my test. I saw them run the same test on the 777 when I was on the program. It did not need chinned tires either. Also the 777 test is in a 777 Boeing video on Certifying the 777.

Dwight Lee Bates

Dwight Lee Bates

Lee's Career as Engineer at Boeing

A guy named Mike wrote:

Lee,
Congratulations on the completion of your homebuilt airplane! I read about it in Sport Aviation and it looks like a fun aircraft. I don't usually write to people out of the blue, but the magazine mentioned something that caught my eye. The magazine article mentioned that you are a retired engineer from Boeing. I've been thinking recently of changing my career and becoming an aeronautical engineer and working for Boeing. I've been an electronics engineer for 26 years in radio, television, and electronic products design and I just earned my FAA airframe and power plant just a week ago. I've been a private pilot for 18 years. I was wondering if you might be able to tell me about the occupation and your thoughts about working for Boeing as an engineer?

Many thanks!

Mike

Mike
I would be happy to based on when I was last at Boeing. I hired a lot of Engineers at Boeing. The first question I would ask is do you do your own car work? Then if they said they graduated from a 4 year Engineering school, I would personally send for a transcript. If both these checked out, I would hire them. You would be surprised how many people lie that do not have 4 year accredited Engineering degrees. Also we would only hire from a good Engineering school like the University of Washington. You have to send in an application on your own. Knowing someone doesn't work due to past problems. Engineering was a good profession for me. I am mechanically inclined and overhauled 10 car engines. I am an

Dwight Lee Bates

Aeronautical and Mechanical Engineer with a Professional Engineering License in Mechanical Engineering. For 44 years, I worked in the automotive, ship building and aircraft industries. I always worked for a Chief Engineer direct which sometimes was a problem due to jealousy of fellow workers. I worked in Boeing in the Golden Age when we were like family before we merged with McDonnell Douglas which changed everything. I took the hard projects no one else would take. You can not become an Aeronautical Engineer with your Electrical Engineering degree and with your A and P. You need a 4 year degree in Aeronautical Engineering. However you might find a job in Electrical Engineering out in the Boeing factory building airplanes. Be prepared to prove you have a 4 year degree from a good college. My ex Boeing boss, who was a Chief Engineer over 600 people, helped me get my Thatcher airplane ready to fly. He came over to Ellensburg from Seattle. I worked for him as Chief Engineer for 13 years. You should try being out in the factory where the fun is. For example, I was promoted into Management but did not like it since I all I did was go to meetings. So I asked to be transferred back to the factory as an MRB Engineer. Also Boeing has cut back on many of the benefits. I was able to take early retirement at age 55 with full medical paid for by Boeing. This is not available anymore so I am glad I took it.

Lee

MRB

MRB stands for Material Review Board. It is a Certification by the FAA at Boeing. The Engineer with this rating can fix any part on any Boeing airplane. It takes two years of working with another Engineer with an MRB Certification. He checks your work and is the deciding factor to your getting the Certification. A mechanic can not fix the part until he gets the disposition from the MRB Engineer. It is a coveted position. Anyone with this rating gets a lot of respect. I got mine in 1996 while working on Composites out in the Everett factory.

Typical Job Description:

Provides Engineering disposition of material, parts, systems, assemblies, and components which do not conform to Engineering drawings, specifications, or quality and manufacturing requirements. Activities include functional test support, systems troubleshooting, generic part replacements and/or designing repairs to restore non-conforming or damaged parts to full structural, functional, and quality capabilities. Respond to Quality Assurance Reports (QARs), Modification Discrepancy Reports (MDRs) and flight Squawks received from the production floor and flight line in a timely manner. Research Technical Orders, drawings, Wiring Diagrams, Vendor Drawings, and other design documents to ensure that dispositions and retest requirement are in compliance with the design documentation.

Dwight Lee Bates

January 25, 1996
B-YS30-DJS-M96-0389

To:
G. Cook	9U-EA
MRSA QC	0W-26
G. Goddard	0X-JC
V. Little	OM-38
W. Woodrow	OX-MX

Subject: Authorized Personnel – Boeing Commercial Airplanes – Materials Review Board – **Production.**

Please add the following names to the subject listing:

Name	Orgn.	Project
Lee Bates	B-YS30	707/727/737/747/757/767/777 KC-135/E-3A/E-4B/E-6

D. J. Serrill
B-Y30 OX-MU
266-7980

Dwight Lee Bates

JOB INTEREST

Engineering

Dwight Lee Bates

OBJECTIVE

Problem solving Engineering position using my 46 years experience. Built Light Sport Aircraft from scratch solving over 200 Engineering problems.

EXPERIENCE

1-6-2012 to 6-30-2013. Composite Quality Engineer, The Boeing Company, Composite Manufacturing Center, Frederickson WA

Boeing called me back to fix composite Engineering problems on the 787 and 777 airplanes. I won 10 awards and wrote 10 Boeing articles how I did it using Root Cause Analysis. I saved Boeing a lot of money at the Composite Manufacturing Center. My total savings is $26 million in my career.

10-1-1999 to 1-6-2012. Engineer, General Aviation, Ellensburg WA Engineering in General Aviation on Light Airplanes since 1999. Built and tested my own Thatcher Light Sport Aircraft from scratch out of 6061 T6 aluminum in 4 1/2 years and 2,700 hours. This is hands on experience which very few people in the world have accomplished. Completed my FAA Private Pilot's License, Tail Wheel Endorsement, Airworthiness Certificate and Repairman's Certificate. At the airport every day working on planes which I love. Built my own website from scratch showing building my own airplane on www.eburg.com/~bateslee/index.html Also did my own stress calculations on my airplane.

9-1-2001 to 12-31-2001 Adjunct Professor Central Washington University Business Department, Ellensburg WA

Taught Advanced Statistics in the Business Department at CWU using my MBA and Quality work experience. Taught the mandatory course entirely in the Computer Lab to hold the student interest. My students loved the class and computers so much that the class average was 93.

Dwight Lee Bates

12-01-1990 to 9-01-1999: Liaison Engineer, Boeing Company, Everett WA

Engineered 747 Final Body Join for Sections 41 and 42 in Manufacturing Process Improvement Group. Ensured the panel loads, assembly and joins were built successfully. Solved problems and reported in the Discrepancy Data Base and to Factory Management. Developed a method to determine causes of rejection tags in the structure for Sections 41 and 42. Engineered and repaired parts in Supplier Liaison with my Material Review Board rating. Had Korean Air Account for Liaison Engineering. Engineered and dispositioned 2,500 rejection tags on aluminum structures and 500 rejection tags on Composite Parts from Northwest Composites Company in Marysville WA. Completed a Boeing Company Course in Tukwila WA on " Advanced Composite Materials Repair for Engineers" which taught me how calculate the stresses and how to repair the Composite parts. Successfully solved 107 problems in late Suspect Discrepancies for a Pride In Excellence Award. Trained in Lean Manufacturing. Served as the only Liaison Chief Engineer's Representative for 3 years on the 777 Project. Saved Boeing $10 million on 777 Sec 41. Completed D19000 Auditors Certificate at Boeing.

5-01-1985 to 12-01-1990: Senior Specialist Engineer, The Boeing Company, Everett WA

Served in Engineering Operations Change Control and Engineering Operations Technical Integration Group improving processes. As Team Leader, my Team wrote a process to put change orders on line on local networks. Another of my teams improved the process flow of Production Revision Records by 16 Manufacturing days. Wrote and presented a training course on Engineering Operations. Won 16 Boeing Pride In Excellence Awards for my excellent performance.

1-01-1978 to 5-01-1985: Department Head Business Review GS13 Department of Defense US Navy, Seattle WA

Managed Industrial Engineering for Cost and Schedules for the overhaul of Navy Ships in private shipyards. Used the Critical Path Method and to track schedule and Cost Performance Reporting to track costs. The 12 ships I Engineered were delivered on the average of 12.4 days early and under budget. Earned four awards for my excellent performance.

2-05-1972 to 3-06-1974: Warn Industries Marketing Manager, Seattle WA

Helped manage a 300 person company which manufactured accessories for the automotive industry. Managed a one million dollar annual national advertising campaign.

Conducted Engineering and Marketing Research to support production line. Won 1972 American Marketing Association Award.

EDUCATION

- 06/01/1971: Seattle University - MBA, Seattle, WA, Major: Marketing

- 02/01/1966: University of Wyoming, Laramie, Wyoming, Bachelor's Degree

Major: Mechanical and Aeronautical Engineering Degree

SKILLS

Overhauling automobile and aircraft engines; Repairing composite and aluminum structures, built and tested my own Thatcher CX4 Light Sport Aircraft from scratch, 40 hours flight testing first Boeing 747. Excellent writer. Wrote a book on poetry and six books on my memoirs. Won 207 Awards.

ADDITIONAL INFORMATION

Licensed Mechanical Engineer State of WA 1978; FAA Licensed Private Pilot since 2002; Ellensburg Experimental Aircraft Association President and Treasurer; Member Yakima Aero Club flying Cessna 152 and 172 airplanes; Member Boeing Incredibles 1969 to present; Calculated stresses in CX4 airplane; Boeing Material Review Board Certification; Ellensburg Environmental Commission for 10 years: Recently Honored as Boeing Incredible at 40th Anniversary of Boeing 747 airplane. Won 10 awards for my 1967 Cougar with 740,000 miles I put on in 45 years. FAA Certified as Composite Material Inspector. I got 100 percent on the Certification exam.

Puget Sound Naval Shipyard

I worked as a Mechanical Engineer at Puget Sound Naval Shipyard from 1974 to 1978. It was good Engineering. I worked on all the different kinds of ships from destroyers, carriers and nuclear "boomer" submarines. I led the work as Planning Yard for 640 Class Submarines. This meant that any improvement or fix I decided to be done had to be done by the other shipyards. This is because after the Thresher disaster, the Navy wanted each submarine in the class to be identical. A new book called: " Blind Man's Bluff" tells all the secrets I knew but could not talk about at the time. The information now has been declassified. The United States submarines used to chase the Russian submarines on patrol all over the world. Sometimes during these chases they would collide. It was a cat and mouse game. All at the same time, the US submarines on patrol surfaced and radioed in the clear to headquarters the position of all the Russian submarines they were chasing. This showed the Russians that we knew the position of everyone of their submarines on patrol. Also they looked for cable warning signs on the shore near the Russian bases. They found one that said : " Do not anchor here due to underwater cables." Then they sent divers out from a submerged submarine and found a communication cable. They then tapped this cable to get the Russian's secret messages they transmitted for 4 years. To fix submarine systems, I went to the Pentagon and Norfolk to read casualty reports. These were reports by the crew on what failed during their last cruise in a submarine. I chose the worst conditions and interviewed the submarine crews on these problems. I asked them about the problems and how they thought that I should fix it. I was one of the first to do this. I then drew up my improvements and fixes and went to the manufacturers. I fixed and improved the systems by working with their Engineers. Then I would follow up by reinterviewing the crews after the changes were made to judge how I did. For example, one crew member suggested using a method which became a Navy standard. These crews were really smart. All the systems I fixed were "Top Secret". At San Diego when I was aboard the carrier Ranger, the Executive Officer called me to his office. He said they were going to deploy in 48 hours on exercises and the main catapult had problems. He was aware of Puget Sound Naval Shipyard's reputation so he asked me to fix it. I called the office at PSNS and we fixed it in 24 hours. I liked most of the Naval officers I worked with. While at Washington DC, we had problems with muggers and Russian spies.

The 50 Percent Review Conference

In 1979, I chose to be assigned to Ship Repair in SUPSHIP because they had a lot of problems. The last ship was 2 months late and millions over budget. I took an office in the Todd Shipyard in Seattle. I immediately saw we needed a scheduling tool. I knew the Polaris Missile development was tracked by PERT (Program Evaluation and Review Technique) and CPM (Critical Path Method). So we wrote into the contract for a monthly submittal of both of these programs by the contractor. On the 50 percent Review Conference of the USS Marvin Shields, I presented a negative float chart showing the main feed pumps were the critical item since it had the largest negative float. This created the largest path in time in the network of the ship's overhaul. Also I made a 5 by 5 foot chart of the network showing the critical path highlighted in yellow. A Navy Admiral sent to straighten us out saw my presentation. After the meeting, he asked a lot of questions about my network. I thought he would take this back to the Pentagon and NAVSEA and I was right. I was called to NAVSEA in Washington DC to show how I did it. I then showed them how we wrote it into the contract. I spent a whole week writing how to do it in a NAVSEA Procedure which became standard for every overhaul in the US Navy. When I got back to Seattle, I was called into the Supervisor of Ship Building's Office. I knew he was going to praise me but I was unprepared for all the Awards he gave me in front of the command in a ceremony. Then I formed my own group with Management Analysts - GS 11s. Each was assigned to a ship to track cost and schedules. We became the best SUPSHIP for over hauling Navy ships in private shipyards. We averaged 13 days early and 12 million dollars total under budget. Then I got requests from other SUPSHIPs across the country to help them implement the contract we wrote for cost and schedules. I was promoted to a GS 12 and given 6 step increases. I flew all over the country in 250 trips by air in 4 years. Then I was promoted to GS 13 Department Head. When I left to go back to Boeing, everyone said I should have been promoted again so I wouldn't leave.

Dwight Lee Bates

Dwight Lee Bates

Marketing

In 1971, I got my MBA from Seattle University in Marketing. My thesis was "An Analysis of Test Marketing Consumer Products in the United States." I love Marketing. In 1972, I took a position as Marketing Research Manager for Warn Industries who manufactures and markets accessories for 4 wheel drive vehicles. My job was to work directly with the Director of Marketing who ran the company of 300 people. I had to help Engineer and market 4 wheel drive vehicle electric winches, hubs, overdrives, differentials, back up alarms, hoists, and do International Marketing. I worked with Engineering to build, test and market the 3000 pound winch. I worked with the Director of Marketing and our advertising agency on a million dollar a year advertising campaign. I won the American Marketing Association Award for a paper I wrote entitled: "Warn's Turning Point." I presented this paper to 100 people at a dinner where I received the award. I demonstrated the Warn 2000 pound electric hoist in Centralia, Washington for Pacific Gas and Electric management. The brake failed so I had an Engineer bring another one up. By the time we replaced it, everyone had left. So on my own initiative, I conducted the demonstration, took pictures, prepared the report and mailed it to the P G and E management. I loved doing International Marketing all over the world. I worked in the Asian market with K. C. Chen, a smart trader. We bought everything under the sun from Mitsui Industries and sold it in the United States. I had my own office and a beautiful secretary named Tonya. I arranged company outings such as roller skating, backpacking and golf tournaments. I wrote articles for magazines, put together press kits, worked with Detroit, worked Trade Shows and did Marketing Research. I put my test data into the sales brochures. I worked at the sales desk and in the repair shop. We made a lot of money on parts. I was responsible for finding new products which meant talking to everyone who had an idea. Also I wrote the Prospectus for Warn going public with its stock. In 1999, I bought a Warn 8000 pound electric winch for a fellow Boeing worker. I got it at dealer cost from the President and CEO Mike Warn. I drove to Portland to pick it up and Mike treated me to a lunch and factory tour. Mike said they were still using my ideas 30 years later. I left Warn to live on our 29 foot sailboat in Port Orchard, Washington and to work as a Design Engineer at Puget Sound Naval Shipyard.

The MiniWinch

In 1972, I was hired at Warn Industries as a Manager in the Marketing and Engineering Departments. I helped develop a new product called the Warn MiniWinch. This small 15 pound electric winch with a 125 to 1 planetary gear box could pull a 3,000 pound direct line pull. The first time we tested the prototype winch I was amazed. We mounted it on a company Jeep and ran the cable to the loading dock. When we operated the little winch, it drug the Jeep across dry pavement with all 4 wheels locked. The Warn MiniWinch now has become famous. It is sold all over the world on All Terrain Vehicles. I even bought one for my own use. I see that competitors have copied the design. They build them overseas and much cheaper. If you want quality though, buy Warn. I was proud to see the MiniWinch had it's own big assembly line when I visited the factory in Oregon in 1999. Mike Warn was still running the company then. I heard that he went to a different plant each day and took Friday off to fly around in his airplanes. He sold me a Warn winch for a Jeep for a friend of mine I worked with at Boeing. After that in my opinion, Mike must have been burned out since he sold the company to a big conglomerate. I glad that I got to manage the MiniWinch development and that it is a success now.

1996

XD9000I WINCH

The XD9000i has a heavy duty Series-Wound Power motor with a winching capacity of 9000 pounds. Also comes complete with a remote control unit, 125 feet of galvanized aircraft cable, clutch, roller fairlead and battery cables. This winch is perfect for those heavy duty self-recoveries!

#27400 **$829.99** each

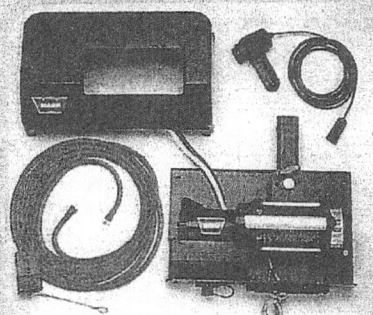

PULL PAK / RECEIVER MOUNT PORTABLE WINCH

Warn's Pull Pak is a sleek, portable winch package incorporating a 3500 pound winch and all electrical connection components. The receiver mount fits easily into a 2" Class III receiver hitch, then folds away for storage. Also comes complete with a remote control unit, 60 feet of wire rope, clutch and automatic in-line brake.

#30500 **$449.99** each

UTILITY WINCH

Warn, the world leader in self-recovery winches, now offers the best value and performance in utility winches! The 1000 and 2000 series winches are the quietest winches in their class with simple operation, corrosion resistant steel mounting bracket, heavy duty electrical powerlead and 30 feet of wire rope. Whether you're winching a watercraft, snowmobile or an ATV, Warn's new winches will get the job done!

#31700 (1000 Lbs. Series) #31702 (2000 Lbs. Series)
$179.99 each   **$199.99** each

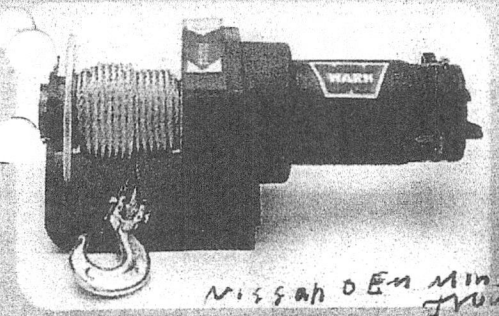

Nissan OEM Mini Truck

ATV Winch

The world's most popular ATV winch! The choice of thousands of ATV owners, this compact workhorse has a 1,500-lb. capacity that takes the back-breaking work out of ATV self-recovery, gathering firewood, farm chores and other tasks. ATVs can get you through a lot but if yours ever gets you into something it can't get you out of, it's reassuring to know you've got this excellent insurance package that'll be able to haul you out of nearly any sticky situation. Also makes it easy as 1-2-3 to haul in heavy loads. Uses the same planetary gears as Warn's heavier-duty winches for trucks and autos. A special feature is the safety interlock that prevents operation of the winch without a key. Clutch allows free-spooling of the 50-foot, 3/16" wire rope. Comes complete with fairlead, mounting plate, remote handle-bar switch, battery cables. Easy to install and use. Don't get stuck without one. Wt. 23 lbs. KE-50610-000..$299.95

Lee Saves Boeing Ten Million Dollars

In 1990, I found that a Boeing division was charging the Boeing 777 Airplane Program ten million dollars to build the Section Verification Vehicle. When we looked at the program budget, I found that the Verification Vehicle was already in the budget. In other words, we had already paid for it and a division was double dipping. I told this to the Chief Engineer and he said it was all right since it is all one Boeing Company. I said it does matter and each division of Boeing was its own profit center. He disagreed. So I found a friend who was retiring in two weeks and had nothing to lose. I sent him to the 777 Quarterly Production Review Conference with a letter I wrote explaining the double dipping of ten million dollars. He presented my letter at the Conference and the division backed down. Later the Chief Engineer called me and wanted to present us with an award for saving the 777 program ten million dollars. I said you disagreed with me and now you want to give us an award? I thought forget about it since I was not listened to in the first place. My Chief Engineer Red McCallum wanted to give me an award for it but I said the award has lost all its meaning by now.

The Resource Center for the Handicapped

For 10 years while at Boeing from 1989 to 1999, I worked as a volunteer at the Resource Center for the Handicapped in Seattle. I helped design the curriculum for Computer Aided Design Associate Degrees for the handicapped. We got accredited from the State of Washington an Associate Degree after one year of study. Boeing then hired some of these handicapped people for helping to design aircraft parts. I mentored students and made sure we could use their skills at Boeing. Also I wrote Boeing newspaper articles on RCH. This was part of my paying back to society. The U S Government paid the handicapped to stay home, not work and be depressed. However, the handicapped wanted to contribute to society. One of the guys I recruited was in a Boeing national TV ad in his wheel chair. I helped talk a friend of mine from the SST and 777 days, Phil Condit who was then CEO and President of Boeing, into passing out diplomas at our graduation ceremonies. When he did, there was not a dry eye in the house. We won the first President Bush's Points of Light Award for the Nation. Our organization and what we did got huge publicity which caused it to be copied all over the country. The President of RCH was Rich Walsh, shown in the following photo, who always said that I was his brother. He used to call and chat with President Bush all the time when I was with him. This showed the wonderful character of this President. He is well loved for his Point of Light Program which honored those who volunteer. I convinced a number of fellow Boeing managers to come to RCH and participate. My boss Red McCallum even gave me an award for my work at RCH. Also my name was on a plaque in the RCH Headquarters as one of the founders of RCH. It is shown on the following page.

...ource Center for the Handicapped
Charter ECAD BAC
April 1991
Loren Lyon
Celerex Corporation
Founding Chairman

Lee Bates	Andy Jeschke
Boeing Airplane Co.	Digital Equip. Corp.
Robert Blankenship	Charles Klabunde
Datasyle	Boeing Advanced Technology
Allyn Boday	Gregory Kott
NUWES	Mentor Graphics
John Boutsikaris	David Nierscher

Dwight Lee Bates

Online Highways Home >Washington > Renton

Resource Center For The Handicapped

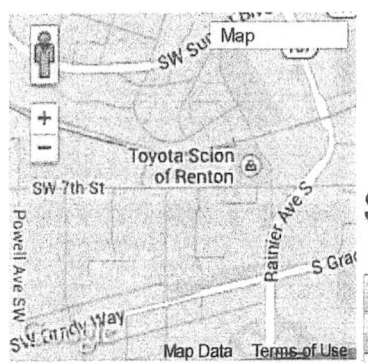

Search Renton Washington

		Search this term or ...
Jobs	Hotels and Motels	Bed and Breakfasts
Airline Tickets	Real Estate	Mortgages	Maps
Weather		

Resource Center For The Handicapped is a non-profit private institution, located in Renton, King County, WA. Total enrollment is approximately 35. There are 5 programs leading to associates degrees.

Popular programs include Computer Programming, Elctmech Instrumentation & Maint Tech, Drafting and Administrative and Secretarial Services.

Location: 500 Sw 7th, Renton Washington 98055 *Telephone* 206-271-0587 Toll Free: 206-430-2241

Dwight Lee Bates

The 1969 Boeing Incredibles

I am telling this story for those who keep asking me what it was like to be a Boeing Incredible in 1969. Boeing Incredibles were those people who built the Boeing 747 in 26 months. We assembled the first 747 in snow storms as they were constructing the building around us The Incredibles slept at their desks rather than to go home. I remember many times leaving to walk out to my car at midnight after working overtime. We were told in a meeting that if we did not FAA Certify the 747, we would lose the Boeing Company. Boeing Management had bet the company when they borrowed so much money that we needed to build, certify, sell and deliver the 747s to remain solvent. The engines were the problem. I flew 40 hours testing the engines for FAA Certification on the number 2 airplane. I wrote most of the 747 Engine Certification Document proving to the FAA that the engines functioned over the flight envelope. I pushed this flight envelope. For example, we had the pilot go to full thrust reverse at 130 knots while landing at Boeing Field. As I expected, the engine surged. The FAA Engineer sitting next to me said: " I realize that you have a problem." The engines would only last 20 hours and we would have to replace them. In fact, we used up 55 engines certifying the 747. I filmed the surge with my 16 millimeter camera. Then I enlarged the frame showing the spectacular surge flame to 8 by 10 photos. I passed out 50 copies at the office until the Chief Engineer said get them all back and destroy them. I did but kept one for myself. At the 40th Anniversary of the 747 at the Museum of Flight, I gave a framed copy of this engine surge to Joe Sutter, the 747 Chief Engineer, after my speech on how bad the first engines were. I remember during a Pre Flight Meeting at the Flight Center at Boeing Field, we stopped the meeting to hear on the PA system the first moon landing. That night while in the air testing the 747, the co pilot came back and told us that the astronauts were walking on the moon. I never saw this until 20 years later. We delivered our number two 747 called Clipper Storm to Pan Am. My pilot friend Dick Pingrey who taught me to fly said he flew this 747 many times. The following photo with me in it taken at Boeing after 20 years shows the 1969 Boeing Incredibles with the 747 airplane we built, tested, certified, and delivered.

The Incredibles of 20 years ago pose with the 747 they helped build
— photo by Richard Green

PLANE BULLETIN

Boeing Commercial Airplanes
Everett Division Employee
Communication, 342-4771, M/S OA-80

BOEING 747

Who Were the Incredibles?

The Incredibles were the Boeing employees who made the 747 fly. From the establishment of the Everett Division in July of 1966, they worked to design the spacious jetliner; transform the remote wooded site to a manufacturing plant that boasts the world's largest building; and assemble and test the worlds largest commercial airplane. The most incredible part is, they did it all in just 26 months.

Twenty years ago this week, September 30, 1968, the world watched as the hangar opened and the first 747 rolled out. The event took place on schedule, thanks to the incredible efforts of the Everett Division workers who had been symbolized by the Paul Bunyan insignia.

Happy 20th Birthday, 747!
September 30, 1988

Special Events

 Birthday Cake: Complimentary birthday cake will be served during lunch (all shifts) on Friday, September 30.

 Rollout Airplane Display: Thursday, September 29, the first 747 will be on display on the apron south of the main factory. A display of incredible 747 history and photos will be found near the airplane.

 Incredibles Photo: A group photo will be taken at 11:30 a.m. on September 29 by the first 747 parked on the apron south of the main factory. Anyone who worked at the plant prior to the first rollout is invited to join in for a group photograph.

777 Design Build Team

In 1990, I was selected by the Chief Liaison Engineer, Red McCallum, to represent him as Chief Engineer on the 777 Design Build Steering Team. It was comprised of 53 knowledgeable Engineers who had previous airplane program design experience. This steering team led the design build teams to design the 777 Boeing Airplane. Team members were at the Chief Engineer level. During our Critical Design Reviews, we invited in passengers, airline pilots, flight attendants and mechanics. We asked them what should we put into the airplane design. For example, an airline mechanic wanted a bigger latch on the engine cowling so he could remove it in a Chicago winter with gloves on. A flight attendant told us how she wanted the cabin lighting. So we put everything in they wanted. We got complete cooperation from the airlines. "Working Together" was our motto." As a result, the 777 airplane is the most loved jetliner by everyone. Now the passengers keep voting it every year as the best jetliner. Our design concept was you fix it on the spot. You do not pass on a problem but fix it now. Do not throw it over the fence. We put the entire airplane on a computer so we could design in 3 dimensions with CATIA. I attended most of the design review and design build team meetings. I would then discuss their problems at the Chief Engineer level. I did this until all the Engineering drawings were complete. Then I moved back up to Everett to work on Buyer Furnished Equipment problems for the Chief Liaison Engineer. The leader of our 777 Design Build Steering Team was promoted to Boeing Vice President after our excellent work. I worked with him when he was at Everett and I started the first Quality Improvement Team there. As a result, he knew the work that I was capable of.

Dwight Lee Bates

In Appreciation

Glenn Aldrich	Rodger Duncan	Wayne Henriksen	Steve Lynn	Glenn Smith
Carl Alu	Debbie Dwyer-Oshiro	Jeanne Henry	Bob Mann	Jack Southall
Gary Anderson	Peter Engquist	Horst Homar	Roy Parkes	Mike Spence
Lee Bates	Phil Farcy	Dale Johnson	Robert Ridgwell	Bill Stonebraker
Mark Breece	Scott Forster	Glenda Johnson	Ted Scoville	Dave Taylor
Kevin Brown	Kevin Fowler	Steve Johnson	George Sevick	Jim Tsai
Neil Christianson	James Garland	Gilbert Key	Henry Shomber	Ken Welever
Pat Copenhaver	Jack Gucker	Charlie Kyle	Joel Short	Andy Wright
Bill Creel	Chris Hall	Steve Lavender	Jim Simek	Bob Young
Chuck Dougherty	Phil Hanna	Ernie Lee	Les Sims	Jim Zumoto
	Gene Haseleu	Stanley Lefever	Rick Skiba	

Thank you for your dedicated support of and participation in the 777 Design/Build Steering Team. This team has been and will continue to be key to the success of the 777 Program. You are turning "The Plan" into a reality.

Working Together to Design and Produce the Preferred Airplane Family.

Henry A. Shomber
Senior Manager,
777 Design/Build

Charlie E. Kyle
Chief Project Engineer,
777 Engineering Operations

John F. Gucker
Chief Project Engineer,
777 Airplane Integration

BOEING 777
Working Together

Dwight Lee Bates

Why I Want to Return to Boeing as an Engineer

12-25-11

Boeing has always wanted me to come back to work since I retired in 1999. They did not want me to retire in the first place since I was a critical skill as an MRB Engineer in 747 Final Assembly. When the offer got to 6 figures and was at the Composite Manufacturing Center in Frederickson, I accepted. Below is listed the reasons why I accepted the offer.

1. Want to make a difference.

2. Got good money offer.

3. I want to fix the airplane.

4. I am one of those left who has a lot of airplane building experience.

5. I love building airplanes.

6. Need some thing to do of a higher calling.

7. I loved building my own airplane from scratch.

8. I was FAA Certified with my MRB Rating.

9. I am a licensed Professional Engineer.

10. I have confidence.

11. I miss the camaraderie with Engineering people.

12. I enjoyed being a hands on trouble shooter for Chief Engineers.

13. Boeing has been good to me with good benefits and 5 good offers and paid for my MBA.

14. I have won many awards in my career and want more.

15. Want to prove a 68 year old can still do it.

16. I am tired of bad press about Boeing.

17. The industry is short of Engineers.

18. I know I can do it due to my track record.

19. I heard the mechanics are stifled in fixing the airplane.

20. Like to work.

21. Proud of my track record on 747, 777, Osprey, 767, 737 and SST.

22. I heard present leaders are not getting out of their offices and out to the factory.

23. We decided on 777 program not to throw it over the fence.

24. Engineering Liaison where I used to work was voted best group in Boeing and were the go between the designer and factory.

25. Have ability to multi task and stay focused.

26. Want to restore Boeing's accountability and respect.

27. Have high work ethic and based on my being a college Professor, I feel few young people now have a high work ethic.

28. Love taking risks in business, flying, mountain climbing and sailing.

29. Love being an Engineer where you have a common goal to build a product.

30. I heard that Boeing Management is pushing fads and not building airplanes.

31. I feel I am ready and must do it.

32. Boeing needs me to come back to work.

33. Boeing builds one of the few products the US still manufactures and has been the US number one exporter for 30 years.

34. Hate out sourcing especially without Pre Award Surveys.

35. Want to bring jobs back to the US from foreign countries.

36. Many future jobs depend on expertise like mine.

37. I want the US and Boeing to build jet liners in the future.

38. I want China, Japan, Europe, Brazil, Canada and India to be behind the US.

39. I believe in gaining market share and not being subsidized like green energy. Let the free market prevail.

40. I have 50 years Engineering experience of which 35 is building airplanes.

41. I know airplane systems, can fix processes, know Quality systems, good communicator, good manager, and analytical.

42. I always ask what does it cost and who is paying for it.

43. Agree with John F. Kennedy who said "Ask not what your Country can do for you but what you can do for your Country." This put men on the moon.

Boeing Composite Manufacturing Center

I interviewed at Everett but no parking and bad traffic caused me not to take a job there. When a Boeing offer to come back to Boeing at Frederickson near Puyallup came in at 6 figures, I took it. It was for fixing composite problems at the Composite Manufacturing Center. They needed my help since I had composite experience. Boeing did not want me to retire in 1999 because I was in 747 Final Assembly in the factory and was a critical skill. Boeing now has to go to India to get Engineers. Our kids do not want to do math. I went to the Middle Schools under the Boeing STEM Program to try to convince them to become Engineers. Everyone at Boeing treated me nice since they heard I was helping them out. I got along especially well with Peter Johnson the Director who was the Head of the Composite Manufacturing Center. He was young, from Australia and an Engineer. When I told him the fix, he got it right away. At CMC, I wrote 10 articles for Boeing publication on how we fixed the composite problems. This would let someone else do it after I left. I was given 10 awards at CMC. One was for $1,500 which I had never seen done before. Every thing I did was Secret including my computer, office and even who I talked to about the problems. I was allowed to talk about the problems only in certain areas of the factory. I had to sign Non Disclosure Agreements saying I would not talk about the composites. Obviously I can not talk about any of these problems in this book. When I solved a number of problems and reretired, they were upset. But they said they would leave my job open in case I wanted to return. People even came up to me in the factory and said that I was the famous guy. I published how I built my Thatcher airplane in a Boeing Magazine. I was on the cover. Many people liked the article. I worked many overtime hours solving problems on the 777 and 787 airplanes. This made building the airplane a lot easier and the factory people thanked me for it. They wanted me to do the next airplane models, but I then reretired. I enjoyed my reretirement party in which every major manager got up to speak about my work. This work led to Boeing Engineers having to do Quality training like I had. We became the poster children of Boeing Engineers since they put us on Quality posters. I saved CMC a lot of money which added to the $26 million total I have saved in my career at Boeing and the Department of Defense.

The Boeing 787 Wins the Collier Trophy

3-13-12

I was proud of my awards and accomplishments on designing, building and repairing the Boeing 787 from 2012 to 2103. We also won the Collier Award for the following Boeing airplane programs I worked on: 747, Osprey and 777. Jim McNerney, the Boeing President and CEO, wrote the following when we won the Collier Award for the 787:

"Today the 787 Dreamliner won the Super Bowl of aerospace - the 2011 Robert J. Collier Trophy for the greatest American aeronautic achievement of the year (2013). What an incredible honor for Boeing, for the thousands of employees who poured their hearts into designing, building and delivering this game changing airplane. Also for the suppliers and customers who contributed to the Dreamliner's success. There is no higher distinction in our field. The National Aeronautic Association has awarded the Collier Trophy since 1911 to recognize innovation in the performance, safety and efficiency of air and space travel. Previous winners include some of the most legendary names in aerospace history: Orville Wright, Chuck Yeager, Neil Armstrong and yes, Boeing. The 747, 767 and 777 have all been recognized with the Collier. Boeing and its partners have also won for the F/A-18E/F, the International Space Station and the V-22 Osprey. It's especially gratifying that this year's award honors the first new airplane of the 21st century. The Dreamliner has pushed the boundaries of flight in countless ways and the Collier Trophy is a testament to the hard work, perseverance and technical achievement of the extraordinary 787 team. Also of their colleagues and partners who helped them and cheered them on. I'm so proud of what you've accomplished. Congratulations on claiming your place in aviation history."

Jim McNerney, President and CEO of Boeing

Jet Liner Bail Out

Mike the jet liners I helped design, build, test and repair at Boeing were never built to use a parachute to bail out. However in 1969, we were trained to use a parachute on the first 747 when FAA Certification Testing. We slid down a fireman's pole to the cargo bay. The cargo bay door with an air deflector would be opened so you could bail out. You had to keep your parachute on or within an arms reach at all times. We had heard stories of a Boeing test that went wrong and the airplane turned upside down. However everyone was pinned to the airplane due to G forces so they could not bail out. This airplane recovered. None of us ever thought we could bail out if something happened. It was an experimental airplane. We realized at Boeing that we had a yaw problem when we swept the wings back. To fix this we put a yaw damper to prevent Dutch Roll. This was never a problem in airline use.

Mike:

This is kind of long but interesting reading. During the 14 years I flew in KC135s we carried parachutes. We strapped on a parachute and put on a helmet during aerial refueling. I did not know of this change:

Subject: Fw: Fwd: RE: [B52KC135] Disaster in the Sky: Old Planes, Inexperienced Pilots and No More Parachutes

Faulty logic on parachutes. Unless the KC135 is stable you could not bailout of the aircraft. If you were stable, then there was no reason to bailout unless you were running out of gas. This was the only case of a successful bailout. This was looked at and decided years ago. In fact it followed a mid-air collision between a KC-135A and an F-4C in August 1976. The KC survived the impact to the rear fuselage, boom and left horizontal and elevator, landing safely at Edwards. The F4C crew ejected and survived. Shortly after, SAC took a look at all KC-135 collisions and accidents, concluding that parachutes were only useful in controlled situations. After that finding, we only carried them on wartime alert.

Telemetry

In 1969 when I was FAA Certifying the engines on the first 747, we sent our data back to the Boeing Flight Center so we could analyze it later. The following photo, taken at the Museum of Flight of the number one 747, shows the engine I certified. Now on the Indy cars they get data from 250 sensors mounted on the car. The Engineers can analyze this data to make performance improvements on the car. Even the TV station broadcasting the Indy 500 can show on the TV screen: miles an hour, RPM and the gear the car is in. Drag racing machines can also do the same thing, The use of telemetry has come a long way since 1969.The following is described in Wikipedia: "Modern telemetry for flight test programs typically monitor data collected from an on board flight test instrumentation over a PCM/RF link. This data is analyzed in real time for safety reasons and to provide feedback to the test pilot. Things that can go wrong are: fading, multipath propagation and the Doppler effect. The bandwidth of the telemetry is often insufficient to transfer all the data required. A limited set is sent to earth for real time processing while an on board recorder ensures the full data set is available for post flight analysis." In 1969 on board the 747 on which I was FAA Certifying the 747 airplane, I had an oscilloscope. This showed the transducer readings to the on board recorder. If I got no readings on the oscilloscope, then something was wrong and we had to stop the test. This happened when the Boeing pilot flew through a cloud. The water plugged the tubing for the flow field rakes so we had to return to the Boeing Flight Test Center to blow out the lines. A friend of mine Addison Pemberton did the instrumentation for the 747 I was testing on. We had 5 airplanes to conduct the FAA Certification of the 747. Later I got to know Addison through the EAA when he built a Boeing Model 40 airplane from plans and a data plate.

Early Retirement

In 1999, I decided to take an early retirement from Boeing. Boeing HR said you are Critical Skill. I was assembling 747s in Everett Final Assembly as a Liaison Engineer. Boeing management tried to talk me out of early retirement. But I retired early anyway. The average Boeing Engineer lives only 7 years after retirement according to a chart I got from Boeing HR. I can believe it. If you are in a high stress job, retire, do not stay active and watch TV all day, you will not last long. I would recommend that if you are one of the lucky few who can get early retirement, then take it. Early retirement and a pension are things of the past. Boeing now puts your money in an account for you. They also match funds which is a real good thing. This means if you put money into your retirement fund, Boeing will also put money into it. But a lot of Boeing people I talked to do not even do this. One guy said I can put the money in my retirement account or take my kids to Disneyland. They spend the money and do not get the Boeing matching fund. So I say are these kids you are taking to Disneyland going to support you when you retire? The answer is obvious. I can not say enough how good the Boeing Company has treated me. I have had 5 job offers from the company and accepted 3. They treat their Engineers really well. They usually pay them good salaries. The only problem I see is see is they do not pay the foreign Engineers enough so they wind up leaving. This is a waste of money. So I recommend early retirement so you will live longer. Start saving your money early and do not touch it.

When I reretired from the Boeing Composite Manufacturing Center in 2013, they kept my job open hoping that I would come back. The CMC Director told me as I was leaving out the door, that he appreciated my work at CMC. He said: "Why are you leaving out the door when you love your job?" I can go back any time I want. They said that you did such a good job on the 787-9 that we want you to do the 787-10 airplane too. Everything fit well on the 787-9. When I retired the first time in 1999, I lost $125K for retiring early since I was a Critical Skill in 747 Final Assembly in the Everett factory. Everyone else who was not a Critical Skill who retired early got a years wages. They had a Human Resources woman follow me around the entire last day. She tried to talk me out of retiring early. They kept sending me jobs to apply for by email. As a 20 year man, I was a level 2 and could never be laid off. I think it was nice to be wanted in 1999 and it was hard to leave in 2013.

Dwight Lee Bates

Early Retirement

Retirement & Life Events

Retirement

- Retirement stages
- Organize your accounts
- Maximize your savings
- Manage your investments
- Define your expectations
- Plan retirement income and expenses

Retirement & life event insights

Retirement, aging and wellness insights from Joe Coughlin, Ph.D., Director of AgeLab at the Massachusetts Institute of Technology.

- What will your lifestyle be in retirement?
- Preparing for your well-being in retirement
- Tips to boost your health and your wellness

Top Retirement Questions

Plan for — or continue to enjoy — a long-lasting retirement.

In retirement
- How can I make my savings last?
- I've lost money — what should I do?
- How can I make sure my income will cover my expenses?
- What are my Required Minimum Distributions?

Near retirement
- When can I retire?
- How can I save more?
- How much will health care cost me?
- Are my family obligations going to affect my retirement?
- What should I do with the company stock in my 401(k)?

Far From
- Where should I save my money? 401(k), IRA?
- How can I save more with all my other obligations?
- What type of investments should I have?
- Should I consolidate my accounts?

Aussies

I love Aussies or Australians. The Director of the Boeing Composite Manufacturing Center where I worked was an Australian Engineer. He is Peter Johnson shown in the following photo on the right. When we would solve a composite problem, we would go directly to his office and tell him how we solved the problem. He got it right away. When I tried to explain it to non Engineers that were managers, they would not get it after I explained it 3 times. Then the follow-up stories I heard the way they told it were wrong. Peter was easy to work with. He told me about the things going on in Boeing Australia. I relayed these thoughts to a friend of mine sent over to Boeing Australia to fix the problems. I am sure Peter will make Vice President which is the next promotional level beyond Director. Also I told Peter this story. At the Navy Department of Defense in Seattle, we were building 2 Frigates which is a ship like a Destroyer for the Australian Navy. Their representatives challenged us to a football game on Sunday. We thought they meant touch football. But what we found out, they meant Rugby. After losing horribly, our guys limped in Monday morning on crutches. The Aussies kept yelling "fair dinkham". I did not know what this meant all these years so I asked Peter. He said it meant:" Are you serious?" The following list of Australian slang words is from kaplaninternational.com. Peter gave me raises, bonuses and many awards. I am sure he will go far at Boeing. Another thing I liked about him is that he always got out of his office and walked the factory floor. Another person I worked with in Boeing Engineering Operations was an Australian girl. She was great. Out of all the countries I worked with in my career, Australia was the best by far.

10 Australian Slang Words You Should Know

Posted by Candida - February 17, 2012 - Fun Facts

Australian slang is almost a language of its own. Aussies (as they're also called) love to play with words, and to use shortened terms to explain things. Though some of them have roots in British English, Australian English has grown and changed into its own interesting type of language.

Many people recognize phrases like "shrimp on the barbie" (which isn't actually said in Australia), and "roo" (for kangaroo), but there are many more phrases that you'd hear during a trip or English course in Australia.

If you enjoy this article, have a look at our Canadian slang and New Zealand slang blog posts! Our English for Teenagers infographic also has many fun slang words.

Here's a list of 10 Australian slang words you should know.

cya this arvo – See you this afternoon.

> *Cya this arvo in class!*

daks – trousers (UK), pants (US)

> *I got some new daks yesterday at the shop.*

dunnie – toilet, bathroom

> *D'ya know where the dunnie is, mate?*

earbashing – constant chatter/talking

> *Her earbashing while I tried to study was driving me crazy!*

fair dinkum – genuine, real

> *Anna's a fair dinkum Aussie*

Meeting a cassowary could ruin your avro.

Dwight Lee Bates

Presentations

A lot of people have been afraid of making presentations. I had a girl that worked for me get up in front of the group we worked with to make a presentation because she said she could not do it. We practiced and she did well. But she said she can not remember doing it. What is it that people can talk just fine in their seat but can not when they stand up? I regard standing up as a way to get peoples' attention. I found out real quick at Boeing that you had to make presentations on your work. If you could not tell people what you did and how important it was, then you were doomed to be unnoticed. The big bosses expected you to be able to make presentations. I was really verbal when I was 5 years old in kindergarten. I practice presentations to make sure I do it right. I have it written down so I can always resort to reading it. I remember after my first wind tunnel test at Boeing. The big bosses not only wanted a written report but they wanted a presentation on the results. I had to get up in front of a group of people to present my work. In those days, people attacked your results so you had to prove you knew what you were talking about. One girl was criticized while she was making a presentation by a manager so bad that she started crying and sat down. Now a days this is not done. I enjoy a big audience. I presented to my EAA Chapter what I did at Boeing after I went back. I got the audience involved or livened them up. I got 25 questions so I judged it successful. The questions are always the hardest since you do not know what is coming and you can not rehearse for them. There are 5 recommended ways to get your audience's attention as I obtained from research:

1. Ask for interaction.

2. Ask a great question at the beginning to get people talking.

3. Get peoples' opinion.

4. Build in audience discussion and reporting.

5. Get moving or get physical by moving around.

CAD

I worked as problem solving Engineer and Manager for 44 years in Aircraft, Ship Building and Automotive fields. I fly my own Light Sport Aircraft which I built from scratch in 2,700 hours and by solving 200 Engineering problems. As a Boeing Everett MRB Engineer, I dispositoned 3,000 rejection tags and was certified by the FAA to Engineer and fix any part on any Boeing airplane. I saved Boeing $10 million on Section 41 for the 777 airplane. I hold an MBA from Seattle University and an BSME from the University of Wyoming. I won 207 Awards in my career. During this period, I used CAD for about 25 years. We put the entire Boeing 777 and 787 airplanes on the CAD Program called CATIA as described following by Wikipedia. Everyone designed their parts on this CAD program. Then all the parts were put together on the computer. If there was an interference, the area would light up. Then you could fix it. Before we used to use a metal mockup at Boeing where you verified that the part fit by physically trying to fit it. I helped develop a CAD Course at the Resource Center for the Handicapped where I worked as a volunteer. Using my curriculum, the disabled students would earn an accredited Associate Degree in 12 months. It was accredited by the State of Washington. These students would often cry when they received their diploma. Some got jobs designing parts at Boeing. I have done design manually also at Puget Sound Naval Shipyard. Our design was drawn with a special pencil on Mylar. This made sure that good copies could be made also the Mylar was easy to erase. Design has come a long way since I first designed in Mechanical Drafting in High School in Cheyenne.

CATIA

From Wikipedia, the free encyclopedia

> This article **appears to be written like an advertisement**. Please help improve it by rewriting promotional content from a neutral point of view and removing any inappropriate external links. *(February 2013)*

CATIA (Computer Aided Three-dimensional Interactive Application) (in English usually pronounced /kəˈtiə/) is a multi-platform CAD/CAM/CAEcommercial software suite developed by the French company Dassault Systèmes. Written in the C++programming language, CATIA is the cornerstone of the Dassault Systèmes product lifecycle managementsoftware suite.

CATIA competes in the high-end CAD/CAM/CAE market with Creo Elements/Pro and NX (Unigraphics).

CATIA

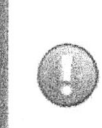

Machine tool simulation

Developer(s)	Dassault Systèmes
Initial release	1977
Stable release	V6R2013x / November 2012
Operating system	Windows / Unix (server)
Type	CAD/CAM/CAE/PLM
License	Proprietary software
Website	www.3ds.com/catia

Contents [hide]

1 History
2 Scope of application
    2.1 Systems engineering
3 Industries
    3.1 Aerospace
    3.2 Automotive
    3.3 Shipbuilding
    3.4 Industrial equipment

Aerospace   [edit]

The Boeing Company used CATIA V3 to develop its 777 airliner and used CATIA V5 for the 787 series aircraft. They have employed the full range of Dassault Systèmes' 3D PLM products – CATIA, DELMIA, andENOVIA LCA – supplemented by Boeing-developed applications.[7]

The development of the Indian Light Combat Aircraft has used CATIA V5.

Chinese Xian JH-7A was the first aircraft developed by CATIA V5 when the design was completed on September 26, 2000.

European aerospace Airbus has used CATIA since 2001.[8]

Canadian aircraft maker Bombardier Aerospace has done all of its aircraft design on CATIA.[9]

The Brazilian aircraft company EMBRAER uses CATIA V4 and V5 to build all airplanes.

Vought Aircraft Industries uses CATIA V4 and V5 to produce its parts.

The Anglo/Italian Helicopter company AgustaWestland uses CATIA V4 and V5 to design their full range of aircraft.

All daughter companies of the French company Safran use CATIA for a full range of aerospace, defence and security products.

The Eurofighter Typhoon has been designed using both CATIA V4 and V5.

The main supplier of helicopters to the U.S Military forces, Sikorsky Aircraft Corp., uses CATIA as well.

Bell Helicopter, the creator of the Bell Boeing V-22 Osprey, has used CATIA V4, V5 and V6.

FLYING

Young Eagles

When I was eight years old, I loved to watch the Stearman crop duster airplanes spray our fields for crop eating bugs. I would marvel at these crop duster pilots flying under telephone lines and pulling back up into the air with the roar of the big engine. Then I got to meet a crop duster called B Mac. He was my hero. He must have noticed me because the next day after he sprayed our crops, he put on an air show for me. He pulled straight up over our farm house hanging the Stearman airplane on the prop. A white scarf billowed out from the back of the cockpit. I was hooked. I started building model airplanes and flying them in my back yard. I only wished that in my day they had Young Eagles. This is when the Ellensburg Experimental Aircraft Association local Chapter 492 flies kids between the ages of 8 to 17 every year. It has been so well received that we fly about 60 kids each year. One year, a smart teacher even brought her entire class to fly in the Ellensburg EAA Young Eagles program. It is a lot of work and expense for the pilots to spend their own money to fly the kids. Some pilots have even flown up from the Tri Cities to fly our Ellensburg kids. I feel the kids of today just want to play video games and need a passion like flying. I have had the passion for flying almost 60 years and recently built my own Experimental airplane from scratch in the shop next to my house in 4 ½ years. And yes I still am building model airplanes. My best moment at Young Eagles came from a young girl who was told she was too young to fly in the Ellensburg Young Eagles. She wrote me a letter telling me why she should fly. The letter said: " I think I should be able to fly because I have gone in my grandpa's airplane. He even let my make it go up and down, left and right and I am going to make an invention so I can fly." She then drew me a picture of a tail dragger airplane. I still have the letter today. Who knows maybe she will become an Aeronautical Engineer like me and design and build airplanes like I did at Boeing? Yes, I let the neat young girl fly with us.

A B-17 Airplane Saved My Life

On a fire in Wyoming in 1964 as crew boss, I was told by the fire leader to take my crew down into a canyon to put out hot spots. I told the inexperienced fire leader that the canyon was a death trap if the wind came up. I had been on 40 fires by then. The fire leader said take your 30 men down there or I will send you home. I put my men in the canyon but climbed to the rim of the canyon to watch for the wind coming up which it did. Then I ran down into the bottom of the canyon and yelled to my men to throw down their tools and run for it. The fire crowned to about 100 foot flames and chased us all the way up to our vehicles. One guy was walking slowly until I told him to look behind him and see the flames. He then ran like hell to the top. At the vehicles, all 30 of my men panicked one after the other. One guy even tried to drive off with people under the truck before we could grab the keys away. We called in on the radio for any fire bomber in the area to drop fire retardant on us as it was our only chance. We were surrounded by fire. A B-17 said he was near us with a full load. We said drop in as soon as possible with out a bird dog guide airplane and gave him directions. Then I yelled for my men to take cover and looked up into the bomb bay as the retardant came out. It hit all of us but no one was hurt. It knocked a gap in the fire and we drove out. You could see all my men in the food line because they had red backs. In 1992, I did the MRB Engineer work on a B-17 I was helping to restore at Boeing. I noticed the stringers in the bomb bay were corroded so I knew it had been a fire bomber. I called the Greybull, Wyoming firebase with the hull number but it was not the B-17 that saved me. They said that B-17 is in a museum in Chino, California. It would have been a good story if the B-17 I was restoring was the same one that saved me. It saved me and now I was saving it. But it was not to be. I put a brick in the Wild Fire Fighter Memorial in Boise, Idaho shown in the following photo from Wikipedia. The memorial brick said: " A B-17 dropped on my crew saving us in the Wheatland, Wyoming Fire in 1964." I own a painting, showing a B-17 dropping fire retardant on a fire, which is hanging up in my shop. It was painted by Rudolph in 2000. The title said: "Taking A Sentimental Journey", a B-17G Flying Fortress operated by Flying Aero Union Corporation of Chino, California from 1960 to 1977. The Boeing B-17 is a wonderful airplane. I have a model hanging above my bed. The following Wikipedia photos show a Boeing B-17 Flying Fortress, a Hercules C130 dropping fire retardant and red fire retardant like dropped on us.

Dwight Lee Bates

Flying Scares

Cessna 152

In the practice area near Yakima, I was practicing emergency landings. I picked a field and turned final. I slowed down to about 50 feet above the ground and decided I could make a safe emergency landing. I applied full power and nothing happened. I was behind the power curve. I slowly got my air speed up and returned to Yakima. Later I heard the area I was practicing in had 100 mph winds. This is what put me in a dangerous situation that I almost did not get out of.

Cessna 172

As I was taking off from Yakima, a big gust hit the airplane. When I rotated to take off, it would not fly. It hung on stall speed until it finally flew off.

TriPacer

As we were flying back from the Oshkosh Air Show, we were approaching Albert Lee, Minnesota to land. Suddenly, the engine quit. The pilot tried everything to restart the engine. We were only 50 feet off the ground and going in. I said: " Switch the tanks." He did and the engine restarted about 10 feet off the ground. We had a fuel leak in the left tank and all our gas had drained out of that tank into the air. The tank we switched to had ½ full tank. The next day we found the fuel leak and I patched it.

Thatcher

On one flight, I made a takeoff run but only got up to 40 mph at full power which was not flying speed. I shut every thing down and coasted to a stop. I later found that the mixture was too rich. I should have gotten 3,000 rpm at run up and should not have tried to takeoff.

Cessna !72

As my brother Scott and I were waiting to take off from Yakima, a huge C-17 did a touch and go in front of us. The tower said:" Clear for takeoff." I said I was going to wait 3 minutes for the vortices from the C-17 to die down. I did and as we took off the plane jumped down and to the right due to vortices. I am glad I waited as long as I did.

Glider Tiger Mountain Down Draft

In 1980, I almost crashed in a glider in a down draft that measured 9 G s on the G meter. We barely pulled out before almost crashing into the tree tops on Tiger Mountain. We took off from the Issaquah Airport.

Bergseth Glider Near Miss

In March 2013, I paid $100 for a glider lesson at Bergseth Airfield near Enumclaw, Washington. A noise in the headset told you if were in a thermal. The louder it got the higher you climbed. I looked up and saw another glider was coming down on us. We were in the same thermal. As I was flying the glider, I dove toward the ground to avoid a collision.

Cessna 152 Near Miss

In 2001, I almost collided with a helicopter near the Pasco Airport. I was landing and had landing clearance from the tower. The helicopter was taking off and said nothing on the common tower frequency. Luckily he dove off to his left and I dove off to my left so we did not collide.

Tail Wheel Endorsement

Since as I understand it, you need to have flown a tail wheel airplane before about 1990 or you need a Tail Wheel Endorsement to legally fly a tail wheel airplane. I had about 100 hours and only flew tricycle airplanes like Cessna 152, 172 and 182 airplanes. As a result in 2009, I signed up with a Tail Wheel Instructor at Cub Crafters with the Sport Cub Light Sport Aircraft. I did fine with my crosswind landings in 30 mph winds. I flew the airplane all the way to the ground and watched the windsock like a hawk. I found if I set up the landing at the right speed and pitch angle, I did good. I also did fine with 3 point landings. But when I went to wheel landings after about 5 hours, I bounced one of them. I had only bounced once before in about 250 landings so when it happened I was surprised and humiliated. I concluded that the reason I bounced was I was not carrying enough power to make the wheel landings. I sat down with my instructor and he agreed. Then due to circumstances beyond my control, the Cub Crafter's Company Sport Cub I was flying left for a nationwide sale promotion tour. I waited for about 4 months and decided to switch to a new instructor and new airplane. This new instructor had 3,000 tail wheel hours and a brand new Citabria airplane. I was afraid I might scratch it and have to wear little white booties. My first flight was to familiarize me with the airplane, so we flew out to the practice area where we did stalls and steep turns. I picked up coordinated turns by stepping on the ball. Then I did my first three point landing and it was like I never left. The next lesson was crosswind landings which I did OK. We were being slammed around by 30 mph gusts which were causing the stall warning to go off all the time. Then every time we tried to fly, the wind would come up just before we got ready to fly. Virgil said I had a dark cloud above my head that followed me around. This happened twice. Then came the day of reckoning - wheel landings again. I remembered that I had to keep the power up and made good wheel landings without bouncing. Then Hurray! Hurray! my instructor said one more good landing of my choice and she would sign me off. I said I would switch over to a three point landing which I did pretty well. As we taxied in, I noticed that I was sweating again which of course I blamed on the airplane's heater. It could not have been my nerves that got sweat all over my instructor's brand new airplane. As we got out, she shook my hand and said congratulations. I finally got my Tail Wheel Endorsement after 5 months and 10 hours flying time. In October, 2009, I could not wait to fly my airplane I built.

Dwight Lee Bates

My First Flight of My Airplane I Built

On my wife's birthday on 4-23-10, I went out to Bowers Field Airport in Ellensburg to fly my experimental LSA Thatcher airplane I built from scratch for the first time. The wind was gusty from 12 to 18 mph and swirling from all directions. I had three EAA members present to keep an eye on me in case I had trouble. My flight instructor Dick Pingrey had flown my plane seven times and 2 hours trying to work all the bugs out of it for me before I flew it. He has 30,000 hours flying planes. After his last flight, we solved the trim problem so he said the plane was ready for me to fly. I high speed taxied out to the end of runway 7 testing the brakes and controllability. Then I ran the engine up to 1700 RPMs for two minutes. Everything was in the green on my instruments so I called out on the radio: "Experimental 27 CX taking off on runway 7." I advanced the throttle so engine torque did not pull me off the runway as my test pilot had advised me. I pulled back on the stick but to my amazement, I was already 20 feet into the air after about 300 feet of roll. I was climbing up at a 30 degree angle at an incredible rate of speed since I now had the throttle wide open. My horsepower is 76 and the airplane was designed for 60 horsepower so I knew I had a rocket. Still I was unprepared for how fast it was. So I shoved the stick forward to stop the incredible rate of climb and looked at my altimeter. I was 100 feet past traffic pattern altitude and I was still over the airport. I looked at my airspeed and I was going 130 mph. I throttled back and entered the traffic pattern over Bowers Field in Ellensburg. I called my position out on the radio. I was overtaking the Cessna 152 ahead of me with the throttle pulled back. So I went way out on my down wind leg to allow the Cessna 152 ahead of me to land. I pulled on the carburetor heat and pulled back the throttle to 1300 RPMs and set forward trim for the landing. I then slowed up to 60 mph and drug it in on final. Then I went to 75 mph on final due to the gusty winds. I was getting hit by big gusts of wind. The bouncing around was a lot more than I had experienced in the Cessna 152s and 172s I had flown. A big gust hit me at the end off the runway knocking me off centerline when I pulled the power off so I banked to the left getting back on centerline. The flare was easy due to the great visibility out of my bubble canopy. My engine has about 10 hours on it and the plane has flown about 2 hours. It is so fast that I have to learn to fly all over again.

Boeing Model 40 and 787

I helped design and build at Boeing the 787 shown in the following photo. I worked on the 787-9 airplane and it went together well and was certified. I worked 500 hours over time. After one and one half years, Boeing Management wanted me to do the 787-10 since I did such a good job on the 787-9 drawings. But I told them I was burnt out and wanted to reretire in 2013 which I did. They left my job open as a Composite Engineer and I can go back any time. They were sorry to see me go. While there, I won many awards for my 787 work solving complex Composite problems. One award was for $1,500 which I had never seen done by Boeing before. I get emails from them. In 1969, Addison Pemberton put in the test instrumentation into the first Boeing 747 test engine I was using to FAA certify the engines. His instrumentation worked great until the Boeing test pilot flew through a cloud which we told him never to do. We had to return to Boeing Field to blow out the lines to the flow field rakes which were plugged with water. Addison duplicated, in the Model 40, flying the mail route across the U S. He showed me a photo of an old mail plane in a 1928 race gassing up at Cheyenne, Wyoming. It later crashed in the fog on Elk Mountain, Wyoming. My Uncle Robert went to the crash and got a strut off the wreck. It was the same strut shown in the following photo. The article with the photo reads: "The pilot, a Lieutenant, following the Union Pacific Railroad and hugging the ground under a low ceiling flew into Elk Mountain West of Cheyenne. The Lieutenant died and his passenger survived with serious injuries." My Uncle said there was nothing left of the wreck so I did not go looking for it. In about 2004, our friend Addison Pemberton invited our EAA Chapter 492 from Ellensburg over to Felts field in Spokane to see his Boeing Model 40 project when it was half done. This is shown in the following photo. He showed us a presentation on his Boeing Model 40. He said Boeing let him look at the plans and take notes. The following photo is from the website - Antique Airfield from the Antique Airplane Association and Air Power Museum in 8-19-14: The 1928 Boeing is the oldest flying Boeing aircraft. It is the first successful airliner in the US. 5339 was delivered to Pacific Air Transport, a division of Boeing Lines, to fly CAM 8 in July of 1928. The airplane crashed October 2, 1928 in southern Oregon when Northbound from Medford to Portland. The airplane was recovered by the Oregon Aviation Historical Society and purchased by Addison in 2000. Credits: Boeing 40C Addison Pemberton/owner. Photo Ryan Pemberton.

Dwight Lee Bates

Dwight Lee Bates

Gyro Copter Flight

About 1993 at the Resource Center for the Handicapped where I volunteered, I was asked by a friend to test fly his new Snow Bird Gyro Copter. He was the founder of Airborne Express and had just bought the Snow Bird Gyro Copter Company. He asked me test it as an Engineer and give him a report. He picked me up in his Lexus car and drove me to the Bremerton Airport. The craft had a 4 cylinder Honda car engine with a pusher propeller. It was 2 place and a flexible drive shaft drove the main rotor for a quicker take off. When I flew it, the main rotor hub bearing overheated and we had to land in a pasture to let it cool down. Also this bearing made a loud noise. My report to the owner and friend was to install a bigger bearing in the main rotor hub. But other than that, I liked it. The gyro copter flew well and sold for about $25K. I told the owner he should sell a kit since $25 k was too much. I found it to be a lot of fun and was worried when gusts of wind hit it. This affected it much different than the Cessnas I was used to flying. I especially thought the craft would be fun to land on a beach with a passenger, have a picnic and to explore the beach area. I would liked to have built one from a kit.

Dwight Lee Bates

Slipstick

A slipstick is a slide rule. As an Engineer that designed and built Boeing jetliners, my pilot friends called me "slipstick." The one I used in college is shown in the following photo. This is before calculators. When the first calculator came out, it was made by Hewlett Packard. In 1961, it's cost was $500 so most college Engineering students used a slide rule. I used one at Boeing after I graduated. When I built my own Thatcher airplane, I named it "Old Slipstick". It is shown in a following photo on the side of the airplane. I photographed my old slide rule. From this, I took it to a Graphics Studio who made it into a decal. I then stuck it on the side of my airplane. The slide rule was hard to use to get the decimal points in the right place. If you moved the center stick past the end, you had to add or subtract a decimal point. In Engineering tests, the decimal point was critical for getting the right answer and a good score. I used to check the decimal point manually. You round off the numbers and do it long hand to check getting the decimal point in the right place. Not many people now a days know what a slide rule was and what it was used for.

C-46

One of my most favorite airplanes is the Curtis C-46 shown in the following photo I took at Paine Field. The airplane is still flying today. It is on the TV show about the Buffalo Air Service in Canada. They fly in the frozen North. Their biggest problem besides crashing in a remote area is heating the engines so that they start. The ruggedness of the airplane was shown on this TV program when they noticed a glow under the cowling while they were flying out in the middle of no where. When they landed, they found the exhaust header pipe had a big hole in it. But it kept flying. A friend of mine used to fly one of these in Alaska. He liked it better than the Douglas C-47. It had bigger engines and could carry a bigger load. He used it for hauling freight. He said he started one with a truck. When the starter broke, they wound a rope around the prop hub. Then when the truck pulled on the rope by driving off, it started the engine. This plane is famous for flying the "Hump" in World War 2. The Japanese cut off the Burma Road so the Americans flew supplies over the Hump. They had to climb to above 15,000 feet and dodge enemy zero fighters to make it. A lot of them did not make it. The route is littered with crashed airplanes.

Jack

When my friend Jack died in 2009, all of us in the Ellensburg Experimental Aircraft Association died a little. He was the glue that held us together. If someone was slow in getting a project done, Jack would get on them. When he died, we had a "Missing Man Fly By" at Bowers Field as shown in the following photo. The other photo shows Jack in the rear cockpit of a Super Cub airplane he helped build from scratch. He was a Professor at Central Washington University in Psychology. I would go over to visit him in his office when I was a Professor at Central Washington University teaching Advanced Statistics. Jack was an excellent pilot. I would often see him early in the morning when no one was around practice landings. Almost all of his close relatives had died from cancer so he probably knew he was next to die. Before he died, he made sure his hangar remained a gathering place for EAA members. He gave me a propeller award for building my Thatcher airplane from scratch. I recited the following poem I wrote at his funeral:

Jack the Pilot

By Lee Bates

11-3-09

There once was an excellent pilot named Jack.

Of all the virtues generosity he did not lack.

He got his Private Pilot's License to fly years ago.

Jack took numerous trips to Oshkosh as we all know.

Jack was always perfect on his landing touch and goes.

You try not to bounce as everyone knows.

Dwight Lee Bates

He was never lacking airplane projects to get into the sky.

Jack had a Cub, a Legal Eagle and Skybolt projects and an RV 3 to fly.

We always borrowed Jack's hangar for our meetings and movie shows.

I even borrowed Jack's shop to finish my airplane from the tail to the nose.

His wife Andrea was understanding and a Saint to behold.

Regardless of his illness, Jack was a lucky man all be told.

We will miss his good nature, kindness and sense of humor.

He was always kidding me and listening to my latest rumor.

Jack is loved by Experimental Aircraft Association Chapter 492.

We will remember Jack as he starts his final flight flying off into the blue.

Dwight Lee Bates

Morgan Marry Me

In 2009, I was at the Ellensburg airport when Mike Butterfield put a banner down on the field to pickup and tow with his Super Cub airplane. The banner is shown in the following photo. He flew low snagging the banner but it would not unfurl so you could read it. I said twist it on the ground so it would unfurl when you snag it. This worked and he flew off to where Morgan was. The guy who wanted to marry her would tell her to look up as the airplane flew by. It worked. The banner said: "Morgan Marry Me." I knew Morgan when she wanted to join my Experimental Aircraft Association Chapter of which I was President. Her mother interviewed me to make sure it was worth joining. Also Morgan was taught to fly by Dick Pingrey who also taught me to fly. In fact as shown in the following photo, Dick gave Morgan a certificate for soloing at the Bonnie Dunbar presentation. I set up the presentation where the astronaut Bonnie showed slides of her 5 space flights. I was glad that I could help Mike Butterfield and I am glad that Morgan said yes. I still see Morgan today since after graduating from Central Washington University, she manages Utopia. This is a coffee and yogurt shop run by Dick Pingrey's daughter and son in law.

Aircraft Mysteries

D B Cooper

In 1971, when I was working in the Warn Industries Headquarters near the Sea Tac Airport, we heard about a hijacking at the airport. After work, I drove over to the North end of the airport. I saw the 727 he hijacked sitting there which was history. He asked for a parachute and money. He then lowered the rear stairs of the 727 while flying to Portland and bailed out. A part of the money was found but he has never been found. Either he died and his skeleton in sitting some where in the river or he escaped. I do not like the entire episode since it started a rash of hijackings all over the world. Since I was in the business of designing, building and testing Boeing jetliners, I did not like the whole thing. My guess is he probably landed in the Columbia River and he is at the bottom of the river. Common sense says that is the only way that some of the money surfaced in the Columbia River. I think people over think a problem like this. I think they want to think that he escaped. However I do not think so. At least I am glad that I got to see the 727 at the Sea Tac Airport. It will always be a mystery.

Amelia

Being a pilot, I have always been interested in the disappearance of Amelia Earhart in an around the world flight in 1937. Lately the disappearance of a Boeing Malaysian Jetliner I helped design and build at Boeing has again aroused my interest. My opinion is that when Amelia could not raise Howland Island where she was to land, that she headed South to the Phoenix Islands. She had a good navigator Fred Noonan who would have recommended this using a line of position. I base this on my experience as a navigator and pilot. To ditch in the ocean would have meant certain death. But to land on an island at least you would have a chance to be seen by your rescuers. I hope one day the people trying to solve the mystery will find one of the engines on her plane. I think it will survive for some time under salt water. I think she landed in the surf and made radio messages until the tide came in. Then I think the plane was carried to the bottom of the ocean by the surf. A way to find it is by a deep sea diving submarine. My friend Dick Pingrey agrees.

From Wikipedia, the free encyclopedia

D. B. Cooper is a media epithet popularly used to refer to an unidentified man who hijacked a Boeing 727 aircraft in the airspace between Portland, Oregon, and Seattle, Washington, on November 24, 1971, extorted $200,000[1] in ransom, and parachuted to an uncertain fate. Despite an extensive manhunt and an ongoing FBI investigation, the perpetrator has never been located or positively identified. The case remains the only unsolved air piracy in American aviation history.[2][3][4]

The suspect purchased his airline ticket using the alias **Dan Cooper**, but due to a news media miscommunication he became known in popular lore as "D. B. Cooper". Hundreds of leads have been pursued in the ensuing years, but no conclusive evidence has ever surfaced regarding Cooper's true identity or whereabouts. Numerous theories of widely varying plausibility have been proposed by experts, reporters, and amateur enthusiasts.[2][5] The discovery of a small cache of ransom bills in 1980 triggered renewed interest but ultimately only deepened the mystery, and the great majority of the ransom remains unrecovered.

While FBI investigators have insisted

D. B. Cooper

A 1972 F.B.I. composite drawing of D. B. Cooper

Other names	Dan Cooper
Occupation	Unknown
Known for	Hijacking a Boeing 727 on November 24, 1971, and parachuting from the plane mid-flight; has never been positively identified or captured.

Northwest Orient Airlines Flight 305

Hijacking summary	
Date	November 24, 1971
Summary	Hijacking
Site	Between Portland, Oregon and Seattle, Washington, USA
Passengers	36 plus hijacker
Crew	6
Injuries (non-fatal)	none known
Fatalities	none known (hijacker's fate unknown)
Survivors	all 42 passengers and crew
Aircraft type	Boeing 727
Operator	Northwest Orient Airlines
Flight origin	Portland International Airport

from the beginning that Cooper probably did not survive his risky jump,[6] the agency maintains an

| Destination | Seattle-Tacoma International Airport |

active case file—which has grown to more than 60 volumes[7]—and continues to solicit creative ideas and new leads from the public. "Maybe a hydrologist can use the latest technology to trace the $5,800 in ransom money found in 1980 to where Cooper landed upstream," suggested Special Agent Larry Carr, leader of the investigation team since 2006. "Or maybe someone just remembers that odd uncle."[6]

Amelia Earhart

From Wikipedia, the free encyclopedia

For other uses, see Amelia Earhart (disambiguation).

Amelia Mary Earhart (/ˈɛərhɑːrt/; July 24, 1897 – disappeared July 2, 1937) was an American aviation pioneer and author.[1][N 1] Earhart was the first female aviator to fly solo across the Atlantic Ocean.[3][N 2] She received the U.S. Distinguished Flying Cross for this record.[5] She set many other records,[2] wrote best-selling books about her flying experiences and was instrumental in the formation of The Ninety-Nines, an organization for female pilots.[6] Earhart joined the faculty of the Purdue University aviation department in 1935 as a visiting faculty member to counsel women on careers and help inspire others with her love for aviation. She was also a member of the National Woman's Party, and an early supporter of the Equal Rights Amendment.[7][8]

During an attempt to make a circumnavigational flight of the globe in 1937 in a Purdue-funded Lockheed Model 10 Electra, Earhart disappeared over the central Pacific Ocean near Howland Island. Fascination with her life, career and disappearance continues to this day.[N 3]

	Amelia Earhart
	Amelia Earhart, c. 1935
Born	July 24, 1897 Atchison, Kansas, U.S.
Disappeared	July 2, 1937 (aged 39) Pacific Ocean, en route to Howland Island
Status	Declared dead *in absentia* January 5, 1939 (aged 41)
Nationality	American
Known for	Many early aviation records, including first woman to fly solo across the Atlantic Ocean.
Spouse(s)	George P. Putnam
Signature	
Website	www.ameliaearhart.com

Missing 777

Based on what I heard on TV and deductive reasoning, I came up with the following possible scenario for the missing Malaysian 777:

1. Takeoff

2. Hijacker holds gun on pilots in cockpit. Locks the door.

3. Copilot is told by hijacker to say to controller "Alright goodnight."

4. Hijacker has pilots turn off ELTs, transponders, ACARS and tells them where to fly to.

5. Pilot puts left turn into flight management system. The auto pilot is on. Pilot's head sets are removed so they can not radio the hijacking.

6. There is a fight for the gun, both pilots are shot. They hit control column in the struggle causing plane to go to 45K feet and down to 30K feet. Holes are shot in the fuselage causing the cabin pressure to escape. Oxygen masks drop for pilots and passengers. Hijacker puts on one of the pilot's oxygen masks. Pilots die from gun shot wounds and lack of oxygen. Crew and passengers can not rush the cockpit unless they have a walk around oxygen bottle. The door is locked and hijacker still has the gun. (Crew at this time should make sure the aft ELT is turned on). The passenger's cell phones will not work since they are too high and no cell phone towers are near.

7. Passenger's and pilot's oxygen (now on the hijacker) run out. Hijacker does not know how to fly the plane to a lower altitude and does not know that his oxygen will shortly run out. Plane is at 30K feet. Everyone suffocates from a lack of oxygen. Everyone is dead now. Plane flies south on auto pilot for 7 hours to Southern Indian Ocean where fuel runs out.

8. Plane crashes in sea and sinks. No ELT signal because they all are turned off.

Dwight Lee Bates

Lee's logic

I think that the gun could have been smuggled on board in a carry on bag. I think that there was no ELT signal from the slides / life rafts since it crashed into a 10 foot high wave and went down immediately. I think the hijacker learned about the 777 systems from googling it on line. I think the Malaysians should have scrambled fighters when it went off course.

I think if the intentional turning the plane to the left at the hand off point to another controller, turning off the transponder, turning off of the ACARS and avoiding Malaysian military shows that the person who did this is very competent. This begs the question why would you want to do this? It is either is a political action to use the plane for political purposes or it is a person who wants to pull this off for personal purposes. We designed the airplane with safety in mind. This missing plane hurts Boeing because we designed it for safety but if a person who is very competent takes over the plane there is little we can design for. The plane is a machine and only does what someone tells it to do.

Dwight Lee Bates credibility for stating the possible scenario is due to the following:

1. Licensed Mechanical and Aeronautical Engineer for 44 years.

2. Has studied all Boeing airplane crashes.

3. Helped design 747, 777 and 787.

4. Had 777 cockpit on his computer at Boeing.

5. Has watched most relevant TV programs since crash announced.

6. Good proven deductive reasoning solving problems at CMC Boeing.

7. Made no mistakes in 1 ½ years at Boeing Composite Manufacturing Center.

8. ELTs, ACARS, and transponders turned off means hijacker.

9. Built my own airplane from scratch and flew it.

10. Tested and FAA Certified engines on first 747 as a Boeing Incredible.

11. 207 awards in career.

12. $26 million saved in career.

13. Solved 10 problems at Boeing Composite Manufacturing Center in last 1 1/2 years.

14. Wrote 12 inches thick memoirs and still going. Secret Clearance Department of Defense on Nuclear boomer submarines.

15. 141 IQ and brain working well. My instincts say hijacked.

16. 777 reliable since 1100 airplanes built, 20 years good flight record, airplane only 9 years old and just was inspected. Redundant airplane.

Lake Cheakamus

In 1992, I flew into Lake Cheakamus, BC in Bob's Cessna 185 floatplane to camp and fish. We flew by the Cheakamus Glacier that feeds Lake Cheakamus. We camped on shore and fished on a raft Dick made. The first photo shows our approach into the lake. The next photo shows our camp on the side of the lake. We flew by the Tusk Mountain which is a volcanic plug that I would like to climb one day. We caught more Trout by a stream running in than we did on Dick's homemade raft. I had hiked to the lake before on my own when we made a trip to Whistler in the summer. The lake is about 10 miles from Whistler. The reason for the green color is the glacial till off the Cheakamus Glacier above the lake. The glacier grinds up the rock as it moves over it about 10 feet per year. This rock powder is flushed down in the lake and gives it the green color. The fish were able to see our bait in the water regardless of the glacial till. We hiked around the lake and found a bed of pine needles. It was like a big mattress you could lay down on and go to sleep. A group of two canoes with 2 young girls in them went by our camp. Bob yelled to them :" Watch out for bears. They like to eat tender young girls." Bob flew 747-400s, a Boeing airplane I helped design and build, to Hong Kong from Vancouver BC. He loved the airplane. I took him to the rollout of number 1000 747 as shown on the cover of this book. Then I took him on a factory tour. He knew about every crash did like I did.

F-18

Boeing builds the F-18 fighter as shown in the following photo by Wikipedia. I wrote the following letter to the President and Congress to promote Boeing's building the airplane in larger quantities.

May 2, 2014

Dear President Obama:

Dear Representative Reichert:

Dear Senator Cantwell:

Dear Senator Murray:

I am writing to request that you support adding Growlers to the FY 15 budget to protect our national security and tens of thousands of jobs around the country.

Currently, the Fiscal Year 2015 President's Budget Request does not include funding for continued production of EA-18G Growlers or F/A-18E/F Super Hornets, both produced on the F/A-18 production line. The Growler is the only full spectrum airborne electronic attack capability, and without your help it would prematurely end production at a time when there is an emerging requirement for additional Growlers to adequately protect military forces in future threat environments.

The Growler is a proven fighter and electronic warfare asset with the unique and essential ability to dominate in any threat environment, making it a critical national asset and needed companion to all combat aircraft. Currently there is no replacement plan or alternate option equal to the Growler. The Navy has submitted an "unfunded priority" to Congress for additional Growlers. If the Growler production line ends, so does the ability to produce electronic attack escorts for the U.S. Navy and joint forces.

In addition to its critical role in battle, Growler production supports highly skilled jobs in 44 states and sustains approximately 800 companies, including many small businesses.

The Growler is a necessary investment for our national and economic security that is proving itself in combat today and is positioned to evolve faster than its adversaries. I urge you to fight to support the addition of Growlers to the FY15 budget to protect our national security and to ensure American manufacturing remains competitive and robust in this industry sector.

Sincerely,

Mr. Dwight Bates

Dwight Lee Bates

Ploesti

When I was a 13 year old boy, I saw a TV program narrated by Walter Cronkite called:" Air Power." It was on every Sunday and I never missed a show. One program that thrilled me was the one on "Ploesti." It showed the bravery of our American flyers who flew from air fields in North Africa over the Mediterranean and mountains to bomb the German held oil fields in Ploesti, Romania. It is why I immediately built a scale model B-24 airplane. The B-24 carried a bigger bomb load and flew further than the B-17 because it had the laminar flow wing. However it was not as well built as the B-17 and was easier to shoot down. In February of this year, an airline pilot giving a talk at our local EAA Chapter said his father flew the Ploesti Raid. Even though no one else in the room knew what we were talking about, I said your father was lucky he did not get shot down. He said his father was shot down and used the resistance forces to escape. The Ploesti Raid was called Operation Tidal Wave. It took place just 18 days before I was born. It is famous because the lead navigator was killed a crash before they got to the target. The replacement navigator made a wrong turn right into the majority of the German flak guns. They were dead meat. But they flew on rather than to abort the mission even though they realized that it was suicide. The source of the photo of the B-24s on fire is: roconsulboston.com in 2014.

Wikipedia:

Operation Tidal Wave was an air attack by bombers of the United States Army Air Forces (USAAF) based in Libya on nine oil refineries around Ploesti, Romania. It took place on 1 August 1943, during World War II. It was a strategic bombing mission and part of the "oil campaign" to deny petroleum-based fuel to the Axis. The mission resulted in "no curtailment of overall product output", and so was deemed unsuccessful. This mission was one of the costliest for the USAAF in the European Theater, with 53 aircraft and 660 aircrew men lost. It was the worst loss ever suffered by the USAAF on a single mission. It was later referred to as "Black Sunday". Five Medals of Honor and numerous Distinguished Service Crosses were awarded to Operation Tidal Wave crew members.

Dwight Lee Bates

Night Flying

I do not like to fly at night. The only time I did was my night cross country flight from Yakima, Wenatchee, Ephrata, Ellensburg and back to Yakima. This was required to get my Private Pilot's License. The thing I did not like was flying over mountains I could not see. You are only judging the fact you will not hit one. Also what do you do if the engine quits? An emergency landing on a strange field with just landing lights is not fun. The following about night flying is from Wikipedia:

a. Remember that the airplane doesn't know that it's dark.
b. On a clear, moonless night, never fly between the tanker's lights.
c. There are certain aircraft sounds that can only be heard at night.
d. If you're going to night fly, it might as well be in the weather so you can double count your exposure to both hazards.
e. Night formation is really an endless series of near misses in equilibrium with each other.
f. You would have to pay a lot of money at a lot of amusement parks and perhaps add a few drugs to get the same blend of psychedelic sensations as a single night weather flight on the wing.

Larry Fookes, a friend of mine, a fellow EAA member and a retired flight instructor for the US Air Force, has the following to say about night flying:

"One night while flying the T-38 (White Rocket) with a student on a night formation flying training mission, the other instructor and myself took the controls and tested each other for a few minutes to show the students real formation flying. At night, there is no horizon to give you a sense of up and down or a sense of where you are in space. The stars are just like the ground lights and if you get disoriented, you don't know which way is up so you must rely and absolutely trust your lead aircraft when you are in the wing position. On this night, the other instructor had me on the wing and proceeded to do some rolls and higher "g" maneuvers. I was somewhat disoriented and then he did a fast roll - what I thought was a 360 degree roll, which puts me on the outside of the circle pulling some "g" forces to keep up. After 180 degrees of roll (being upside down in my mind as compared to the earth), he stopped and rolled back upright. I had the sense that we were going to continue the roll so it came as a surprise and really messed with my mind and I got a case of vertigo that I will never forget as now I was essentially going up hill on the outside of the circle while having to input controls forward, which was a downhill type movement. It took everything I had mentally to stay on his wing and not break away. I was so messed up in my mind that breaking away would have been very dangerous as I would not have known where I was in space, upside down, vertical, etc. or where he was in space so I would not, once free, run into him. It was so intense, that it is a memory that will stay with me even though I get Alzheimer's. You learn to trust what you know and not what you feel."

Plane Crashes on Elk Mountain

I called Uncle Fred to see if he knew anything about the strut Uncle Robert had from the crash of the Standard Biplane in the air race about 1928 I am writing about in my Memoirs. He said he did not remember that crash but remembered the crash of the DC-3 in 1946. He said: It happened in January. About 7 ranchers, Fred and other people from Elk Mountain went up to the crash site to retrieve the bodies. He said Henry Chesbrough was one of the ranchers. They drove up the exposed ridge on Elk Mountain with their pickups until blocked by snow drifts. They then walked to the crash site with the sled dogs. They concluded the pilot saw the mountain top and pulled back on the stick but the tail of the DC-3 hit Elk Mountain. He said they found all but the pilot from the 21 on board. The bodies were in good shape since it did not burn. The plane was in 2 pieces as the tail broke off. One of the engines flew clear over the ridge and down the other side of Elk Mountain. They slid the frozen bodies off the steep south side of Elk Mountain. There at the bottom, the dog teams picked the bodies up. Then later they went back to the crash site to find the pilot. They dug in the snow and found him. He was thrown clear of the wreckage. The authorities did not show up for a long time to process the bodies. Fred knew the ranchers hauled off the metal to the scrap yard. He said at least 3 airplanes have crashed on Elk Mountain. In later years, he went up to the crash but found very few pieces left. He thought people like me and the ranchers had picked it clean.

I found a website www3.gendisasters.com for airplane crashes in which a guy whose father died in this DC-3 crash was trying to get closure by visiting the crash site:

Comments:

My dad was killed in this crash

Submitted on 19 August 2013 - 8:03pm.

I am taking a trip next week to Elk Mountain to be where the plane crashed ending my dad's life. I was only 7 months old and I am now 68 years old. But need to finally get closure on saying goodbye to the dad I never knew! I plan on leaving him a letter on the mountain top.

So I posted the following on the website:

On 9/9/2014 8:28 AM, Lee Bates wrote:

I read your trip report on the crash of a DC -3 on Elk Mountain in 1946. I found the crash site twice in 1960. It was to the West of the lookout by two huge broken off trees. I first went to it with a friend of mine I hayed with and later my cousin Rod. I think they would have not crashed if they had not hit the two big trees. Each tree was about 3 foot in diameter. They were only about 50 feet from the top. Most of the parts were just beyond the 2 trees to the South East. I found a DC- 3 access cover to the gas caps there verifying it as a DC -3. When I showed it to my Uncle Robert, he also identified it. He said: " I recognize the access cover to the gas caps since when I have flown on a DC -3, I have seen those access covers flopping in the wind when they forget to close them." I heard that a rancher took his pickup to the top and hauled the metal to the scrap yard to sell. This is probably why the crash site is hard to find today. Also I heard they used the electric starting motors to the DC-3s engines as electric winches for their trucks. I am trying to find out information on a Standard Biplane that crashed into Elk Mountain in the 20s in a race. My Uncle Robert had a wooden strut from it which I identified from a photo of the Standard gassing up in Cheyenne during the race. The article with the picture said the pilot died and his passenger had serious injuries. It was heading West in the air race after gassing up. They were following the Union Pacific Railroad tracks, hugging the ground in a low ceiling. I think Sherman who tried to help you find the DC -3 crash site is my cousin. Please call me. The guy called me and said he visited the crash site of the DC-3 on Elk Mountain in 2013 to get closure. He was only 7 months old when his father was killed in the crash. He said a rancher with ATVs took him to the site. He said they used a road made by the University of Wyoming to access their cloud seeding building on the top of Elk Mountain. He said that he took a picture of the crash site shown in the following photo but there was not much metal left. I said there was not much metal also left in 1960 when I visited the crash site. I can verify that this is the crash site since I can see the two broken off trees that they hit in the distance in the photo. Also I remember this is what the crash site looked like when I found it twice. I told him I had just talked to my Uncle Fred that day who helped bring the bodies down after the crash. I said he is 88 years old so I think he is the only one left who helped bring the bodies down. I said the propeller hub and 2 propeller blades were at the gas station by the Hanna cutoff on the main freeway for years that came from the crash. I used to see it all the time when I stopped for gas.

The following explanation of the crash is from:coloradowreckchasing.com.

Elk Mountain DC-3 Crash Site Background:

The Crash : January 31, 1946 United Flight 14, a DC-3 aircraft en route from Salt Lake City to Cheyenne, Wyoming at an assigned altitude of 11,000 feet crashed on Elk Mountain, Wyoming. There were 21 fatalities.

The accident report described a following United flight observing UA 14 diverging slightly south on the airway east leg from Sinclair. The trailing flight also reported an orange flash in the dark about 3 miles to the south. That was the end of UA 14. The wreckage was found on a direct line between Sinclair (SIR) radio range and the Laramie radio range (LAR). The approved airway for the area was actually a dogleg connecting the east leg of Sinclair to the north leg of Laramie in order to skirt around the 11,152' bulk of Elk Mountain. The accident report did not outright state it but the conclusion this author draws is that the flight did not follow the dogleg range legs but, instead, used the ADF to track directly from Sinclair to Laramie. This would have worked out if the flight had been at 13,000' as originally filed rather than at 11,000' as requested en route. The impact site was at 10,700 MSL according to the report (more like 10,900 by altimeter and GPS). In computer simulation conducted, it seems that there is a 30 second window after passing SIR (Sinclair beacon) where taking a course for LAR(Laramie beacon) results in ground contact. Turning earlier or turning later winds up missing the mountain.

The following website gives more history of crashes into Elk Mountain: www3.gendisasters.com: One of the early mail planes came to grief on its miles-long, tree covered slopes, deep in winter with snow, and a private pilot reportedly at one time crashed there. Two Army planes crashed into the mountain within a month of one another in January and February of 1942. Another bomber with a lone pilot rammed the mountain a month later and the wreckage was not found for some months. In June 1930, the mountain area was the scene of the biggest search in Wyoming history. For two weeks, hundreds of searchers combed the Elk Mountain region for two boys, 4 and 6, separated from their parents on a fishing trip. They were never found.

DC-3 Crash Site Photos

The following photos are DC-3 crash sites. The tail section is a crash of a DC-3 in Canada which killed 7 when it flew into a mountain. The other photo is on Elk Mountain and is where I found the DC-3 wreckage. It was taken by the guy who lost his father in the crash who called me. He said he got closure by visiting the crash site.

Dwight Lee Bates

CARS

Dwight Lee Bates

1967 Cougar

1967 Cougar on KIMA TV 11-15-10

My 1967 Cougar and I were on KIMA TV. Attached are blurred photos off my TV. Basically the story went. Lee Bates, a retired Boeing Engineer, who built and tested the original 747 has a way to beat the recession. He has driven his 1967 for 43 years and has 718,000 miles on it. He also has life time parts on it. He takes a broken part into a parts house and gets a free one. The parts houses thought Lee would only keep his car 5 years. But he has gotten 9 free mufflers and 9 free batteries. Lee says he is a cheap skate but his wife just says he is frugal. He even built his own Light Sport Airplane in 2,700 hours and 4 1/2 years. He loves his car and it is easy to work on. Lee says why not keep a car if it is a good car? He has driven his car equivalent to 30 times around the earth or to the moon and back and to the moon again. At this rate, he will get a million miles in 17 years.

Lee's 1967 Cougar Facts

11-15-10

1. Bought the Cougar in 1968 for $2,200 in Burien with 30,000 miles on it. It was the Motor Trend Car of the Year in 1967. 2. Has 718,000 miles on it now. I have driven it almost every day for 43 years. 3. Is on its' 6th engine. 4. I put in two engines myself. 5. Has original C-4 transmission which has been overhauled 1 time. 6. Gets 15 MPG in town and 20 on highway 7. Had the most miles on a Cougar at the World Cougar / Mustang Rally at Bellevue Community College and according to the Ford Motor Company. 8. Ford gave me an award for it at 500,000 miles. 9. Car and owner have been in Ellensburg Daily Record Newspaper twice and on Yakima TV KIMA twice and KOMO TV and radio once. 10. I took out 289 engine and put in 302 engine so it would burn regular gas. I have always had a V-8 engine and love them. 11. I take it out in the desert where jeeps go and it handles just fine. 12. Fun car to drive. 13. At 3,400 pounds and with a V-8 it is perfect weight and power for

me.   14. I changed out the differential twice, upholstered the interior 3 times, painted it 4 times, replaced the vinyl top 2 times. 15. Car has many lifetime parts which I take out, take down to the parts house, get another free part and put it back in myself. 16. The only major problem was when an engine broke in the Oregon desert. 17. Everywhere I go people ask me about the car. I have them guess how many miles it has on it. No one has been close. 18. All the miles and repairs are documented in my car books.   19. License plate reads 1967 XR7. 20. I disconnected the headlight doors since they do not build the vacuum motors anymore. 21. I disconnected the flip away steering wheel to save wear on it. 22. The car will probably be given to a museum when I die. 23. It has been wrecked 9 times but the original metal is still in place since it can be easily straightened. 24. I drive it every day and have done so for 43 years. 25. I sleep in it when I go out in the desert prospecting. I take the seat back off on the passenger side and put in a board and foam mattress.   26. I would like Ford to use it in its advertising if I can get a hold of my old friend Alan Mullaly the President of CEO of Ford.

Cougar

First generation (1967–1970) [edit]

The introduction of the Cougar finally gave Mercury its own "pony car". Slotted between the Ford Mustang and the Ford Thunderbird, the Cougar would be the performance icon and eventually the icon for the Mercury name for several decades. The Cougar was available in two models (base and XR-7) and only came in one body style (a two-door hardtop). Engine choices ranged from the 200 hp (149 kW) 289 in^3 two-barrel V8 to the 335 hp (250 kW) 390 in^3 four-barrel V8. A notable performance package called the GT was available on both the base and XR-7 Cougars. This included the 390 in^3 V8, as well as a performance handling package and other performance enhancements.

The 1967 Cougar, with the internal code T-7, went on sale September 30, 1966.[3] It was based on the 1967 refaced first-generation Mustang,[4] but with a 3-inch-longer (76 mm) wheelbase and new sheet metal. A full-width divided grille with hidden headlamps and vertical bars defined the front fascia—it was sometimes called the electric shaver grille. At the rear, a similar treatment saw the license plate surrounded on both sides with vertically slatted grillework concealing taillights (with sequential turn signals), a styling touch taken from the Thunderbird. A deliberate effort was made to give the car a more "European" flavor than the Mustang, at least to American buyers' eyes. Aside from the base model and the luxurious **XR-7**, only one performance package was available for either model: the sporty **GT**. The XR-7 model brought a simulated wood-grained dashboard with a full set of black-faced competition instruments and toggle switches, an overhead console, a T-type center automatic transmission shifter (if equipped with the optional Merc-O-Matic transmission), and leather/vinyl upholstery.

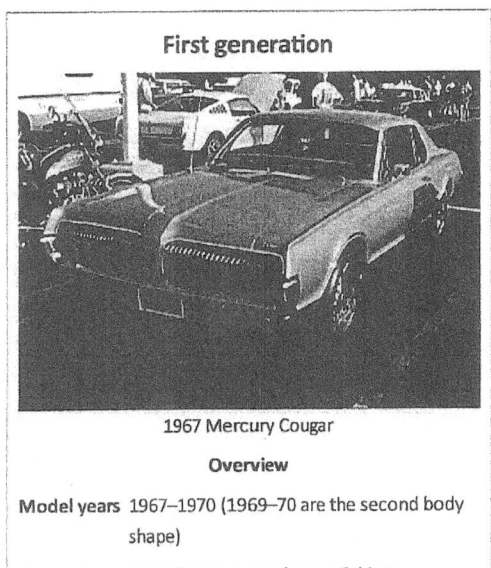

First generation

1967 Mercury Cougar

Overview	
Model years	1967–1970 (1969–70 are the second body shape)
Assembly	United States: Dearborn, Michigan
Body and chassis	
Body style	2-door hardtop coupe 2-door convertible
Layout	FR layout
Related	Ford Mustang
Powertrain	
Engine	289 cu in (4.7 L) Windsor V8 390 cu in (6.4 L) FE V8 302 cu in (4.9 L) Windsor V8 351 cu in (5.8 L) Windsor V8 428 cu in (7.0 L) FE V8 427 cu in (7.0 L) FE V8
Dimensions	
Wheelbase	111 in (2819 mm)

The GT package, meanwhile, supplied a much larger engine, Ford's 390-in^3 (6.4 L) FE-series big block to replace the small-block 289-in^3 (4.7 L) standard powerplant. Along with this came an upgraded suspension to handle the extra weight of the big engine and give better handling, more powerful brakes, better tires and a low-restriction exhaust system. Introduced with the music of Herb Alpert and the Tijuana Brass' *The Work Song*, the Cougar was a sales success from its introduction and helped the Lincoln-Mercury Division's 1967 sales figures substantially. The Cougar was *Motor Trend* magazine's Car of the Year for 1967

Dwight Lee Bates

Lee on 1967 Cougar on KOMO Radio the Ken Schram Show on November 15, 2010 at 2:45 PM

On KOMO Radio Monday at 2:45 PM from Seattle I was interviewed by Ken Schram for 11 minutes on my 1967 Cougar car. The show went like the following:

Ken: The guy we are going to talk to is a man after my own heart. He has driven his 1967 Cougar he bought in 1968 for 43 years. He bought it when he was just starting out at Boeing as an Engineer designing, building and testing the first 747. He researched the car which was Motor Trend Magazine "Car of the Year for 1967." What is it like to drive this Cougar for so long? You are my hero.

Lee: Thanks for inviting me. You are my hero for the TV program "Town Meeting." I can sum up my feelings with a poem I wrote OK?

My 1967 Cougar

By Lee Bates

3-12-09

In 1968, I bought a beautiful used 1967 Cougar car.

Little did I know it was destined to become a star.

I replaced each engine at about 150,000 miles.

Every day my mileage guessing game brings smiles.

Dwight Lee Bates

No one has come close to winning the contest.

One day a guess of 500,000 miles was the closest.

The truth be known it is 718,000 miles actual.

The number of engines is six and is factual.

I laugh at the huge 30,000 dollar new car cost.

The fact I tell people I love my car is not lost.

When I head into the desert, in the Cougar Camper I sleep.

In the mountains, it crosses creeks that run deep.

As a cheap skate, I like the fact it has lifetime parts.

The fact that I do not pay for them is good for us old farts.

I also laugh at the car payments I never had to pay.

When I die, the Cougar will go to a museum to stay.

Ken: Hold it a minute for the applause since the whole studio is applauding and laughing. I want to thank you for giving your time and I really admire you and your 1967 Cougar. I have an old car myself. Tell us about your lifetime parts.

Lee: I got 9 free Lifetime mufflers and 9 free Lifetime batteries. When a part breaks, I take it off, take it to the parts house and get a free part and install it. Ken if you are ever in

Ellensburg give me a call and I will give you a ride. Thanks again for having me.

Ken: Thank you for a great interview.

<u>Lee on TV and Radio</u>

11-15-10

No.	Show	Date

1. Condo Association Meeting Seminar TV Seattle 1978

2. Ski Adventure Show Coliseum TV Seattle 1995

3. 1967 Cougar Car Story KIMA TV Yakima 2000

4. Sasquatch Story KIMA TV Roxanne V. Yakima 2005

5. 1927 Robin Airplanes Story KIMA TV Roxanne V. Yakima 2006

6. 1967 Cougar Car Story KIMA TV David Klugh Yakima 2010

7. 1967 Cougar Car Story KOMO TV Dan Lewis Seattle 2010

8. 1967 Cougar Car Story KOMO Radio Ken Schram Seattle 2010

Engine Swap

In 1996, I needed to put new engines in my 1967 Cougar and 1966 Mustang. A local engine remanufacturer had a discount for Boeing SPEEA employees which is a Professional Aerospace Union I belonged to. So I ordered two engines - one for my 1966 Mustang and one for my 1967 Cougar. The cost was $800 per engine. One weekend I put in the Mustang engine and the next weekend I put in the Cougar engine. I had both long blocks delivered to my house at the same time. The following photos show only the Cougar engine being swapped. I took off all the engine accessories off the old block and put them on the new block before I put the new engine in. The first photo shows the old engine in the Cougar. The second photo shows taking the accessories off the old block and putting them on the new block. The intake manifold and carburetor are obvious accessories. The photos show the old engine being hoisted out and the accessories changed. I used a rented engine hoist and my own engine stand. The reason I put both engines in myself is I got a good price for both engines and I knew I could do the job right. I used a Shop Manual for each engine and checked off each step. This is important so you do not forget any thing. I have overhauled 10 engines in the past, but now I just put in a long block. It is far easier than overhauling an engine. My engines have lasted from 250,000 miles for a Ford Crate engine to 130,000 miles for an engine bought from an engine remanufacturer. The important thing is that they put in all new parts and give you a 50,000 mile warranty. Most engine remanufacturers put heat plugs on the block. This way they will void the warranty if you over heat the engine. If you do, the plugs fall off. I enjoy this kind of work. I also put the engine in my Thatcher airplane when I built it from scratch.

51 Chevy

When I worked for the Bureau of Land Management in Craig, Colorado, a guy I worked with asked me to help him tow his 51 Chevy to the dealer. He was trading it in on a new car and it did not matter that it did not run. So I switched the wires around before we towed it to the dealers. Then since I did not have a car at the time, I went back to the dealer 2 days later. I said how much for the 51 Chevy? The mechanic said he could not get it to run. I said I like 51 Chevys so I will give you $25 for it. The mechanic said get it out of my sight. We towed it back to the BLM Compound where I was living right out of the dealer's with the same car we towed it in with. He should have suspected something. I switched the wires back where they were supposed to be. It started right up. I went to the junk yard and got a used battery for it. I ran this car for 2 years and 25,000 miles while at college. I built a model of the car as shown in the following photo I took. It had a door I put on to replace the door damaged in a wreck. However I could not get it to open or close so I welded it shut. I called it my 3 door since only 3 doors worked. At college, I took a date out in it. She told me I had a nice car because I only let her in and out on the good side. She never saw the wrecked side. As I was driving out to Seattle to work at Boeing, it threw a rod. It was near Kemmerer, Wyoming in the middle of a snow storm. So I pulled out my sleeping bag and went to sleep. In the morning, there was 2 feet of snow on the ground. I saw the only truck in 4 hours coming so I got out in front of him. He had to stop or hit me. He gave me a ride into Kemmerer where I caught the train to Seattle. I called a friend I knew from Kemmerer and asked him to tow the car to the junk yard. I said keep the money for your efforts. Later at Boeing, he sent me a check for $50 he got from the junk yard. I made $25 on the car.

Dwight Lee Bates

Racing

Formula One

I like to watch Formula 1 car racing on TV because I like the cars and what it takes to be a winning driver. I wrote a story about a race car driver called Senna in these memoirs. He is acknowledged to be one of the best before he was tragically killed. The RPMs of the engines is fantastic. The engines are revving about 15,000 RPMs at max which is amazing. Also 500 HP used to be high. Now it is 1000 HP. A lot of the cars have carbon fiber or composite bodies. These parts are easily replaceable. But the problem is the cost. I have built Boeing jetliners out of this composite material and I know how much it costs to build these parts. I heard on TV that is costs $30,000 to repair composite parts on a Le Mans type race car. I think making the tub which protects the driver out of composites is a good idea for safety. No one has been killed since Senna. I like to see the technology that goes into these cars. I had to sign a lot of non disclosure statements at Boeing because I know so much about building and repairing composite parts. My most favorite Formula 1 race is Monaco which has many curves and no place to bail out. I loved the movies " Grand Prix" and "Le Mans" which shows how the cars and safety has improved. But it is a rich man's sport and I can only watch on TV. Maybe one day when my oil money comes in, I will have a front row seat at Monaco.

Senna

I was impressed by a documentary on the Brazilian race car driver Senna. In 1994, his car was bad and he knew it could kill him but he did not quit. He went straight into the wall with such an impact that everyone knew his was dead. He died of a head injury. I have always liked Formula 1 racing. The following photo showing Senna racing is from Wikipedia. I feel that Senna was beloved in Brazil his country and was a great driver. He held 3 World Championships for Formula 1. In my opinion and according to the TV program, bureaucracy killed him. They had the Formula 1 race cars equipped with sensors and a computer to control the car from skidding. The traction was automatically adjusted by the computer.

This made it very safe. Then the bureaucrats eliminated this feature. Even a good driver like him could not control an unbalanced car with out the computers help. Since the fatal Senna crash, they reimplimented the computer and there has not been a death since before I wrote this. This makes Senna a hero in my mind. He was good looking too. The women went nuts over him.

Indy

As I write this, I have just watched the Indy 500 in 2014 where Hunter Reay won it by .06 seconds or 10 feet. The strategy as I see it is:

1. Try to be 4th up until near the end to save on fuel burn. This makes the last fuel pit stop shorter on time. Let the leader who is pushing air burn more fuel.

2. At the end, try to take the down force out of the wing to be able to go faster. Do not take too much down force out or you will spin.

3. Have quick pit stops.

4. You have to be in top 3 in last 5 laps to have a chance to win.

5. Be in the first 2 in the last lap.

6. Pass on the last lap to take the lead. This prevents the guy behind you from passing you on the last lap.

7. Do not go 3 wide in the turn as this will make someone crash.

When I saw Rick Mears win in person at Indy 1979, the wind tunnel and aerodynamics were coming into play with the ground effect cars. The aerodynamics is so refined that now a tunnel fairing is behind both rear tires. The attendance in 2014 was only 200K instead of 250K when I attended. Rick Mears whose autograph is shown on the following page has won 4 times. Carbon Fiber is becoming the material of choice for the Indy cars. I have listened to the Indy 500 for the last 55 years. It continues to thrill me.

Formula One

From Wikipedia, the free encyclopedia

"F1" redirects here. For other uses, see F1 (disambiguation) and Formula One (disambiguation).

Formula One, also known as **Formula 1** or **F1** and referred to officially as the **FIA Formula One World Championship**,[2] is the highest class of single-seater auto racing sanctioned by the Fédération Internationale de l'Automobile (FIA). The "formula", designated in the name, refers to a set of rules with which all participants' cars must comply.[3] The F1 season consists of a series of races, known as Grands Prix (from French, originally meaning great prizes), held throughout the world on purpose-built circuits and public roads. The results of each race are evaluated using a points system to determine two annual World Championships, one for the drivers and one for the constructors. The racing drivers, constructor teams, track officials, organisers, and circuits are required to be holders of valid Super Licences, the highest class of racing licence issued by the FIA.[4]

Formula One cars are the fastest road course racing cars in the world, owing to very high cornering speeds achieved through the generation of large amounts of aerodynamic downforce. Formula One cars race at speeds of up to 350 km/h (220 mph) with engines currently limited in performance to a maximum of 15,000 RPM. The cars are capable of lateral acceleration in excess of five g in corners. The performance of the cars is very

Formula One

Category	Single seater
Country	International
Inaugural season	1950[1]
Drivers	22
Teams	11
Constructors	11
Engine suppliers	Ferrari · Mercedes ·Renault
Tyre suppliers	Pirelli
Drivers' champion	Sebastian Vettel (Red Bull Racing)
Constructors' champion	Red Bull Racing
Official website	www.formula1.com

Current season

Formula One

Current season	[show]
Related articles	[show]
Lists	[show]
Records	[show]
Organisations	[show]

VTE

dependent onelectronics – although traction control and other driving aids have been banned since 2008 – and on aerodynamics, suspension and tyres. The formula has radically evolved and changed through the history of the sport.

While Europe is the sport's traditional base, and hosts about half of each year's races, the sport's scope has expanded significantly during recent years and an increasing number of Grands Prix are held on other continents. F1 had a total global television audience of 527 million people during the course of the 2010 season.[5]

Grand Prix racing began in 1906 and became the most popular type internationally in the second half of the twentieth century. The Formula One Group is the legal holder of the commercial rights.[6] With annual spending totalling billions of US dollars, Formula One's economic effect and creation of jobs is significant, and its financial and political battles are widely reported. Its high profile and popularity have created a major merchandising environment, which has resulted in great investments from sponsors and budgets in the hundreds of millions for the constructors. Since 2000 the sport's spiraling expenditures have forced several teams, including manufacturers' works teams, into bankruptcy. Others have been bought out by companies wanting to establish a presence within the sport, which strictly limits the number of participant teams.

PERSONAL

Our Wedding

We were married on August 19th 1967 on my birthday. We had rehearsal in the West Seattle Church and a talk with the preacher. Basically he said to respect each other. The church was beautiful with a forested landscape and flowers behind huge glass windows as shown in the following photo. I arrived early and helped carry in the cake. The wedding went off without a hitch. I did something different than I had seen before at other weddings. I said my wedding vows loud enough so that the whole audience could hear them. In the reception line afterwards, people said that they liked that. My best friends were the ushers. They were: Dick, Fred and Gary Vieth. My Best man was Dave a Navy officer friend I met in my Queen Anne Apartment building. I talked to Gary Vieth when I was at Boeing before I retired when I worked on a 737 electrical Suspect Discrepancy with him. I also called him when I went back to Boeing at CMC. He has worked there since 1966 or 47 years. He still does not want to retire. He loves Boeing. My grandfather came by train to the wedding from Denver. He caught the bouquet. It hit him in the back of the neck and he grabbed it. Torrey Gamlin the head of the Boeing Auburn Fabrication Plant was there. The head of the same plant would give me my Composite Manufacturing Award 46 years later. Diane's parents had a neat party afterwards but we left early. We went to San Francisco for our honeymoon in our 1964 Chevelle convertible. We went to Fisherman's Wharf and the Golden Gate Park which we revisited in 2013. We went to Finocchio's Night Club where gay people dressed up in women's clothes. We drove down the coast to Pebble Beach, Carmel and Big Sur. We ate at the famous Nepenthe Restaurant overlooking the ocean at Big Sur. We walked the beach at Pebble Beach and went to Carmel's expensive shops. I sneaked on the Cypress Point Golf Course to see the famous 16th hole. I had to hide in a bush as a security guard tried to run me off. Diane knew the area from summer trips when she was a little girl. We both love San Francisco but it is too expensive to live there. My brother Jay lives in San Mateo not too far away. Now where we used to live in a condo on Capitol Hill in Seattle is more expensive than San Francisco. The next week I reported back to work on the Boeing Supersonic Transport. We are still married after 47 years.

Dwight Lee Bates

Our House

Our View

I have always lived with a good view. Our view in Ellensburg is a view of the Cascade Mountains and Mount Stuart. We sit up on a hill and can see the Kittitas Valley to the West. There is nothing like looking at the view while you are eating lunch and dinner. The sunsets are spectacular. I like to watch a sunset or full moon out my bedroom window. This is one reason why we selected this house when we retired. I like the view we have of Wilson Creek and our grove of Quaking Aspen trees. Sometimes the sunset or sunrise lights up the Stuart Range with Alpenglow. This is a pink light that causes the snow on the peaks to glow pink. It is the time to take a picture according to Galen Rowell the late photographer. I can see the mansion built on the hillside across the valley by a retired Microsoft Executive. I would have to pay millions for a view like this in California. Also it is cheap living. I can tell the difference between what I had to pay when I was working again for Boeing and now. Our view at Montlake in Seattle was beautiful. We could see the University of Washington and the surrounding tree lined hill sides. I could see the following mountains I have climbed from our roof in Montlake: the Brothers, Pilchuck, Rainier and Three Fingers. Every time I walk by our windows in Ellensburg I can not help but look out our view. As I write this, my wife Diane is telling me about the water level in Wilson Creek below. Also in my bedroom, I can watch all the sunsets. When I see a good one, I run upstairs to tell Diane. You can see it better from our upper level. Also at night when the moon comes up, I open my bedroom blinds to watch the landscape lit up by the moon. I look at our Quakie grove, creek and sunsets every day out our view. I still take the time to look at it since I never get tired of it. At Boeing, I worked in a closed factory with no windows for a year and a half. It was dark when I went to work and when I went home. I worked 500 hours overtime so all I did was work and sleep. Now as I am writing this, I am looking at our view and Mount Stuart as shown in the following photo.

Our Yard

In 1999 when we moved to Ellensburg, we redid the yard. Like we did in Montlake, we had a professional tell us what plants we could put in that matched the climate as shown in the following photo. I also cut the yard in grass down to size so it would be easy to maintain. In Montlake, I made a kidney shaped lawn and in Ellensburg I made a circular shape. Since Seattle got 21 inches of rain and Ellensburg only 9 inches, the plants are different. People who have visited like our yard shown in the following photo I took. Diane loves to work in the yard. We planted Ginkgo Trees in planter boxes. We put in an arbor that you can sit in. Bloomtime Nursery in Ellensburg advises and sells us our plants:

Bloomtime Brings Beauty To Life (from Bloomtime Sales Brochure):
At Bloomtime Nursery, we are open year round to serve your nursery and design needs. We are conveniently located in Ellensburg, Washington and just a quick trip from Yakima.
We offer the following and more:
- Complete Nursery and Design
- Fresh Floral
- Unique Gift Shop
- Fresh Flowers

We can customize our services to fit your nursery, design, and floral needs. We also offer great ideas for holidays and special events.

My Shop

My shop is my escape area. It is where I built my airplane from scratch 3 hours a day for 4 ½ years as shown in the following photos. My shop was built as a high school project so it is well built. It is heavily insulated. Best of all it has a view of Wilson Creek and the Kittitas Valley. I have my favorite pictures and awards on the wall. It is heated for Winter use. Diane seldom goes in it, but I keep it clean. I have music available but mainly enjoy the peace and quiet. I have about 30 model airplanes I built hanging from the ceiling. This is where I organize all my repairs to my cars and house. I lay out the tools, new parts and old parts out on the table I built. I keep the old parts there until I am sure the new parts will work. Most of my tools are organized there. Every man needs a shop.

Dwight Lee Bates

Dwight Lee Bates

Dwight Lee Bates

Dwight Lee Bates

MBA

In 1972, I graduated with an MBA. It stands for Master of Business Administration degree. I went to night school at Seattle University. I took 21 courses to finish. I did the last part as 4 courses, 4 courses and 3 courses. I think writing my Thesis was the most fun. The Albers MBA program was the first night business school program in the country. This enabled people to work during the day and go to school at night. I did this while certifying the first 747 engines at Boeing which was a lot of work. The Professional Master of Business Administration (MBA) program at Albers uses an innovative approach to teaching that integrates skills and concepts into an experiential learning environment. You not only learn valuable business skills, but are challenged to incorporate them in real world situations. If you want to participate in a leading part-time MBA program that is unique in the Northwest, this is the one for you. Developed with input from business leaders and students, the Albers Professional MBA curriculum is streamlined, customized, integrated, connected, and flexible.

Dwight Lee Bates

- **Call (206) 973-8126 to speak to an advisor.**
- Join us at an information session to learn more about the Albers Professional MBA, finance accounting graduate programs.
- Download a viewbook.
- Arrange to visit a class by contacting an Albers Graduate Student Ambassador.
- Apply now.

Request Information

The Future is Here, the Future is Now

Develop your business skills and advance your career at a leading part-time MBA program that is unique in the Northwest and located in the heart of Seattle. Created with input from business leaders and students, the Albers Professional MBA curriculum is streamlined, customized, integra connected, and flexible.

What Sets the Professional MBA at Seattle University Apart?

- **Streamlined:** Earn your MBA part-time in as little as 9 quarters and full-time in 6 quarters.
- **Customized:** Tailor your program to fit your specific interests and career goals.
- **Integrated:** Learn how to integrate your skills and knowledge across disciplines.
- **Connected:** Build lasting connections by taking your core classes with the same group of students.
- **Flexible:** Strike the right balance among your professional, educational, and personal goals.

Roughly 80% of the students at Seattle University's Albers School of Business and Economics wor full-time while pursuing an MBA or other graduate degree. With rolling enrollment (start any quarter), evening classes, and convenient locations in Seattle and Bellevue, our Professional MBA program was designed for them...and you.

To learn more about our Professional MBA program, click here.

For a complete list of Albers graduate programs, click here.

Dwight Lee Bates

The Professor

In the Fall quarter of 2000 at Central Washington University in Ellensburg, I was interviewed for a job as Professor. It was in the Business School and I would teach Advanced Statistics. I thought it would be fun. I had to send the Dean my resume, and grade transcripts from the University of Wyoming for my Engineering degree and from Seattle University for my MBA. I passed the interview. The first day of class, I put the "t test " up on the board and said this is a tool we will use to test our hypotheses. It doesn't make sense but it works so just use it. I quickly found that the students were not interested in lectures. They wanted to be entertained. They did not buy the book because it was too expensive at $100. They would not use the tutors I hired. But I found that they loved computers. So I arranged with the Assistant Dean to get the Computer Lab full time. That was the key. I taught by doing problems on the computer. Everyone had their own computer with SPSS on it. This was a computer software program that did the calculations for you. I said at Boeing you would be given a program like SPSS to use so this is real world. They loved it. If someone that I could see from the back of the room did not have the right computer screen up, I stopped the class. I got them straightened out and we moved on. Even the people next to them were helping. We took all tests on the computer. At the end, the class average was 93. I let my graduate assistant teach also. I had 5 really good students of 35 total who I said should major in statistics. My graduate student was a great help in recording and grading the tests. I had my own office over looking the beautiful fountains on campus. The assistant Dean was really helpful. I rarely saw the Dean. They said that he was in meetings all the time. I talked to the students about Steelhead fishing on the Snake River. Three of them said lets go there fishing. I said that was a conflict of interest. What if you 3 guys got As and the rest of the class thought it was because you went fishing with me? I liked teaching but thought it was too easy to do for a living. Working for Boeing was a lot harder.

Dwight Lee Bates

SPSS

From Wikipedia, the free encyclopedia
Jump to: navigation, search
For other uses, see SPSS (disambiguation).

IBM SPSS

Logo icon, ver. 17

SPSS v.19 x86 running on Windows 7 x86

Developer(s)	IBM Corporation
Initial release	1968
Stable release	22.0 [1] August 13, 2013
Operating system	Windows, zLinux, Linux / UNIX & Mac
Platform	Java
Type	Statistical analysis, Data Mining, Text Analytics, Data Collection, Collaboration & Deployment
License	Proprietary software
Website	www.ibm.com/software/analytics/spss/

SPSS Statistics is a software package used for statistical analysis. Long produced by SPSS Inc., it was acquired by IBM in 2009. The current versions (2014) are officially named **IBM SPSS**

Dreams

As I read stories that other people wrote, I wanted to do the following:

1. Bomb Tokyo in the Doolittle Raid in World War 2. This was due the book and movie:"30 Seconds Over Tokyo." The poster shown is from Wikipedia.

2. Go on the Lewis and Clark Expedition. This was from reading the Log kept on the expedition.

3. Be the first to climb Mount Everest. I followed the climbing stories and was amazed when Hillary climbed it.

4. Be an Ace fighter pilot in World War 2. I read lots of stories on this. I met Gabby the Ace fighter pilot who had the most kills in Europe in World War 2.

5. First to fly the Atlantic Ocean like Lindbergh in his plane the Spirit of St. Louis. The following photo shows a replica of The Spirit of St. Louis airplane I took at the Oshkosh Air Show.

Dwight Lee Bates

Viet Nam

Lee's Comments on Viet Nam

From 1961 to 1966, I went to the University of Wyoming and took Army ROTC. This was during the Viet Nam War. Once at Gene's Bar in Laramie, Wyoming, I bet $30 that I would stand in formation for ROTC. I was a Junior so I was through with ROTC. I stood in formation and the rest of the troops marched off. Since I won the bet by now, I slunk off to the men's rest room to escape. A regular Army officer saw me and yelled: " Hey you, get back in formation." I took off running through the men's rest room with this Army Officer chasing me. His tie was flopping as he chased me at full gallop across Prexies Pasture. I escaped through the Engineering Building. The guy who paid me the $30 for the bet said: "I did not think you would do it." When I got a position with Boeing, I got an Occupational Deferment from Viet Nam. When in college, I had a Student Deferment for 5 years. All of a sudden in 1968, I got a notice from my Draft Board that I was 1-A or fit for the draft. My brother Jay was also turned 1-A by the same Draft Board. They were getting desperate for recruits for Viet Nam. I saw a TV show called :" A Week with Charlie Company" where a truck was blown up by the Viet Cong with American soldiers inside. Viet Nam was a dangerous place. I did not want to die. I thought about moving to Vancouver, BC in Canada and working as an Engineer. However, I went to 3 levels of Boeing management before they agreed to put me on a Top Secret Fighter Program and sent a letter to my Draft Board. Mean while, I had to go down to the Army Depot in Seattle and get my physical with hundreds of young 18 year old kids. Later I concluded that at least ¼ of these kids were killed. I passed my physical. Then I stupidly took the Officers Test where I got the highest grade they had ever seen. I was Officer Material. By this time, they had killed about 30,000 Americans in Viet Nam. Then due to the Boeing letter, I got my Occupational Deferment back. I had just gotten married and did not want to go to Viet Nam as cannon fodder. I was overjoyed. Then later they resorted to a lottery for the draft. I was out of it now due to my age but my lottery number was 152 out of 365 so I would have been drafted. I left Boeing and worked on Navy carriers like the USS Enterprise and USS Ranger that were being used in Viet Nam. So I did my Viet Nam duty as an Engineer. Now that I look back on it 36 years later, I was glad I never went to Viet Nam. The full scope of it hit me when I visited a traveling Viet Nam War

Dwight Lee Bates

Memorial Wall with over 40,000 names on it of kids who gave their lives. I would have gone if forced to go.

Brother Jay's Comments on Viet Nam

I had a student deferment in college and got an Occupational Deferment afterwards, although I was 1- A for a short while and took the physical at Fort Douglas in Salt Lake City. I got all the questions right on their IQ test, except one, and got an Occupational Deferment shortly thereafter. I was offered commissions in both the Army and Marines and never had to decide to take a commission. One of my classmates in Civil Engineering dropped out school and took a commission in the Army and was killed in a fire fight. His name was Gilbert Bush from Laramie. No one from Ault was killed in Viet Nam but a number from Cheyenne were. Still Cheyenne was a good draft board to have as they had plenty to choose from and getting a deferment there was not that difficult. When I lived and worked in Washington, DC, one of my friends had been in Nam as a Second Louie and had a grenade tossed into his tent to warn him to go easy on the pot heads in his command. He eventually went to work for the Senate Public Works Committee. Life is short and I would have gone to Nam if I had to. I never did any drugs, only too much beer.

Brother Scott's Comments on Viet Nam

I had a Student Deferment and didn't want to go to Viet Nam as I had already had two friends killed over there. Just about everyone that I met in college that had been over there and made it back was screwed up on alcohol, drugs or mentally unstable. Dad's best advice to me was: " You don't want to go to Viet Nam and you need to get a college education." It was the best advice I have ever gotten in my life and I followed it. My lottery number was 172 and I would have gone if I had dropped out or flunked out of school. They were drafting up to 242 in the lottery numbers. I made it five years which is the number of college years allowed for a Student Deferment. I needed one more quarter of college to get my degree when I got a notice in the mail in November. It said that I had lost my Student Deferment after 5 years and was being reclassified as 1-A and eligible to be drafted. Lucky for me, President Nixon did away with the draft that January or I could have been drafted. I could been one of the last to be killed in Viet Nam. I jumped up and down in the room for joy when President Nixon announced that he was doing away with the draft. It was a war that we didn't win and a lot of good young people's lives were wasted for nothing. The 60's became a time of rebellion for a lot of young people against the Viet Nam War, patriotism, parents and society. A lot of young of young people experimented with drugs, sex and dropped out of society and became labeled as hippies. I was caught temporarily caught up in the war protests and took part in a peace march at the University of Oregon in Eugene. I marched down the streets of the University of Oregon with an estimated crowd of over 18,000 people. My room mates, which I answered their room mates wanted ad on the University Want Ads Board, were members of the college campus SRA organization. It was a Communist organization that supported Mao Tse Sung. My room mates would pass out leaflets and had a Communist Party table in the student union. The FBI was monitoring our telephone conversations because one time I picked up the phone in our apartment to use it after one of my roommates had just talked. My friends had gone off the deep end.

Uncle Fred's Comments on World War 2

My Uncle Fred went through 2 close calls like this during World War 2. In 1944 during World War 2 as an 18 year old, Fred was drafted. He was assigned to the U S Army in the Signal Corp. as a typist. This was after Basic Training since he could type 122 words per minute. He got to be a good typist when he studied at a one room school house on the ranch near Elk Mountain. The school district bought him a brand new typewriter which our Aunt Marguerite taught him to type on. This was before she married Uncle Donald and she was teaching also. This later saved Fred's life. He was sent to Camp Crowder, Missouri and met Adrian who was to become his future brother in law. He was Eileen's brother. He was forced to drill in the cold rain without a shirt and caught pneumonia which again saved his life. Then, he was sent to Fort Livingston, Louisiana for basic training. His 45th Infantry Division was sent to combat. He was sent later to Scotland and then to Dachau, Germany. He was a typist for the Signal Corp and never saw any combat. His 45th Division was wiped out since most of them were killed in combat. So the pneumonia and being a typist saved his life. This is about the time of the Battle of the Bulge but Fred never fought in it. Maybe Ma knew that he was over there at the time and came up with the Battle of the Bulge story. At Dachau, he saw the concentration camps. He was 4 hours on and 4 hours off as a guard.

He said they kicked the German people out of their homes to be able to sleep in their beds. He did not think that it was right. I said that I thought that the German people knew about the concentration camps even though they said they did not. Then Fred's division was supposed to go to Japan after the war in Europe ended. Next he was shipped to Texas to a typing pool. His buddies were shipped to the Pacific. He typed for the Army and was discharged.

Dwight Lee Bates

Aging

At this time as I am writing this, I am 70 years old. The following photos show me at 4, 24, 47, and 70 years old. It is about 20 years between each of the following photos. I complained to people at work saying when I was 24, I was good looking. I said at 68 years old I am showing my age. They said I still look good. Then I showed the photo of me at 24 years old. They said it was not me. All said that but one girl who said it was me based on the eyes. Why can not I still look like I am 24? I guess the body grows old but the mind does not. Lets see where was I? Oh yes, the worst thing about getting old is the aches and pains. Also we remember the past like it used to be. What do the good old boys talk about when they are over the hill? You guessed it - women. They probably dream about what they used to do. The next thing they talk about is politics. Just let us run the world for a while and we will straighten it out. I said when I was 24 that when I get to be 70, I will not be like those old gessers. Guess what, I am one of those old gessers. Not that anybody listens to me anymore. The problem when you are talking to the good old boys is you can not get a word in edgewise. I have to raise my hand to get attention. This starts them laughing so they do not pay attention to you all over again.

Dwight Lee Bates

Dwight Lee Bates

Les Miserable's

In May of 2014, we went to a great musical play at the McConnell Building at Central Washington University in Ellensburg, Washington where we live. The musical only cost $12 for Seniors. The musical is based on the novel by Victor Hugo. The production was by the Central Theatre Ensemble which collaborates with many local groups. It is to offer the CWU campus and community an unusually diverse selection of dramatic, musical and dance productions for all ages. It grew out of an outstanding academic program in theatre arts that is known regionally for its hands on experimental education. Starting as Freshman, students have the opportunity to participate in all aspects of theatre production, and to experience many unique and innovative programs. For example, CWU's Department of Theatre Arts has the only musical theatre program in Washington State. It has the only Bachelor of Fine Arts Theatre Degree in Washington State. The play story line as I perceived it goes like this. In the 1800s, a man is thrown in prison for 19 years for stealing a loaf of bread to feed his sister's children. This is about the time of the French Revolution. The story is similar to : "The Tale of Two Cities" by Charles Dickens. The man is released from prison. He violates his parole and runs off to another city. He become successful in business and becomes mayor of the town. When he interferes with his workers abusing a pregnant girl worker, he is discovered by the sheriff as the fugitive who violated his parole. In a sword fight, he could have killed the sheriff but lets him go. He helps the abused girl to the hospital where she has the child called Cosette and dies. The man adopts Cosette as his own daughter. Then she grows up and the French Revolution occurs. The patriots form a barricade to revolt against the establishment. Cosette's lover is wounded. The sheriff is captured but again let go again by the man he thinks is a criminal for violating his parole. This criminal man carries the wounded Cosette's lover off to restore him to his health. The sheriff can not figure out how the criminal let him go twice and in despair commits suicide. Later all the people at the barricade are killed. The criminal restored the future husband back to health. He then tells him that he has to flee because he does not want to let Cosette know he was a criminal. Then the husband of Cosette tells Cosette that he was a criminal and they go find him. Then he lives with them until he dies. The moral of the story is the kindness overrides the cruelty. This is sung at the end. The poster shown is from Wikipedia.

Why I Write Poetry

4-26-09

Poetry tells me how to communicate with myself. In compresses my thoughts into a neat package which I can communicate to other people. It comes from my heart. I can express my love for other people, sports and life's challenges. I like the rhythm of the words in a poem. I can express life's meaning to other people. It tells people who I am. It tells others how I feel. It reaches out. It expresses what it means to be a person. It is not about showing off. It is how I feel about it. I am growing old and want to express my life experiences. It tells my thoughts to other people. It might inspire a young person to follow in my foot steps. I am proud of what I did in life and want to tell others how I enjoy life. It expresses the reason for living. I do not want to die without telling others how I feel. Poetry is similar to my love of the arts like music, art, acting and writing. I feel I am intelligent and witty enough with a good memory so I can combine these into a poem. It takes a while but the words come out in the end to make a point. A finished poem is like hitting the sweet spot when you drive a golf ball. You make such perfect contact that you do not even feel the impact. It was like adjusting the carburetor on the engine of the airplane I built. I know when I find the sweet spot and the engine purrs like a kitten. The light bulb goes on in my head when I know I have written a good poem. It is similar to solving a major problem out in the factory on a Boeing airplane when you get that special feeling. You know you have solved the problem. Writing poetry is like flying. I get in such a focused state that I can feel things are going smoothly. Poetry expresses my emotions. I like the rhythm and balancing the number of words on a line. It tells people what is interesting to me. I have to tell people things that are in the back of my mind. I feel my poems are written to my standards and do not have to be perfect. However I do ask people if they like my poems and listen to their feedback. But if I like them that is the way they will be. I write poetry to express my feeling about my life. Maybe I can inspire others. I will be cremated and will not take up any space in this world when I am gone. As a result I hope my poetry book may remind people of my life and how I lived it.

The Ellensburg Address

By Lee Bates

4-29-09

One hundred days ago our voters brought forth on this continent, a new nation, conceived in injustice, and dedicated to the proposition that all poor men who do not work can take the achiever's money. Now we are engaged in a great debate, testing whether that liberals, or the latest administration so conceived and so dedicated, can take money from those who earned it and give it to those who do not work. We are met on a great battle-field of that idea. We have come to dedicate a portion of that argument, as a final resting place for those who here gave their hard work so that this nation might live. It is altogether fitting and proper that we should do this. But, in a larger sense, we can not dedicate - we can not condone it - we can not allow - this Socialism to happen. The hard working men and women, who struggled here in our great country, have consecrated it, far above our poor power to add or detract. The world will little note, nor long remember what we say here, but it can never forget what they did here. It is for us the living hard working people made this country. Let it be they who are dedicated here to the unfinished work which they who have worked so hard to get ahead in this country and have thus far so nobly advanced. It is rather for us to be here dedicated to the great task remaining before us - that from these liberal government programs to take away our hard earned money that we take increased devotion to that cause for which the hard working people in this country gave the last full measure of hard work - that we here highly resolve that these hard working people who earned their money shall not have worked in vain - that this nation, under God, shall have a new birth of freedom from Socialism - and that government of the people, by the people, for the people, shall not perish from the earth by being changed into a Socialist State by the liberals.

Dwight Lee Bates

Grandpa's Poems

I only saw my grandpa on my mother's side a couple of times. Now that I am older, wish I was nicer to him. When I was about 4, I kicked him in the shins with my Jean Boots because he was spitting tobacco everywhere. He hardly had any teeth. I used to wonder how he ate a steak? The following are his poems that we have saved. He wrote the first poem when he was a ranch hand for the Queally Land and Livestock Company Stock Raising in 1900 in Medicine Bow, Wyoming. He wrote his poem from his own experiences just like I did for my poem book I wrote. As far as my research shows, grandpa Jay never copyrighted this poem. Since he is dead, I think he would be happy that I published his poem for all to enjoy. The other 2 poems, on World War 1 and the prostitute, I am publishing my grandfather Jay must have liked since he had them in his possession. They show his character. He was a tough old guy. When he was dying, all the relatives gathered in his hospital room to pay their last respects. He stuck his head out of the oxygen tent and said; "I do not see why you are gathering around. I am not dead yet." They said you were not a man in those days unless you could roll a Bull Durham Cigarette in one hand while riding a wild bronc full speed into a Wyoming blizzard. I used to go to the homestead ranch that he built. The food was kept in a spring the house was built over. They were real men in those days. The best thing the US Government did was the Homestead Acts in 1862 and 1909. They did it to develop the West. If you developed 160 acres and lived on it, you could own it. This gave a lot of immigrants a chance for a new life in the United States. Many came from Europe. To get anything, you had to work for it. I wish the kids of today had to work for things instead of being given it by their parents. Carolyn Johnson bought the old homestead ranch and divided it up among her relatives. We used to stay with them in the old ranch house. The following photo of my brother Jay and I on a sled was taken there. I guess I was 4 and Jay was 6. Note the building in the photo is made of logs probably cut on the property. My great grandfather on this grandpa's side was shot in 1902 in Medicine Bow, Wyoming. It was a gun fight in which many shots were fired and they reloaded. My grandfather evidently drew first but missed. The guy who shot him was tried for murder but got off on self defense. The gun fight was over a dance hall girl. Our family tried to cover it up by saying it was over a card game and not a dance hall girl.

Queally Land and Live Stock Company Stock Raising

<u>The Tale The Bottle Told</u>
By Jay L. Johnson

A drink. No thank you pard
Though to refuse comes pretty hard
For I have been in the toils of Demon Rum
And to answer no bothers me some

I will tell you a story, this a tale a bottle told
Of an old range pal, who has passed into the fold
We were riders, and he and I
Were punching cows for the lazy Y

The boys all called him Sunny Jim
I go by the name of Rawhide Slim
When we all got peeved, sore and riled
He took things cool and I joked and smiled

Out on the round-up when it rained a spell
And we all rolled out at the daylight yell
Grumbling and cussing a puncher's life
Jim would be cheerful mid all the strife

But Jim must have his periodical
And that no doubt made him a prodigal
For all of us boys could tell by his ways
That in his past he had seen better days

After the fall round-up and the beef were in
Winter settled down and it snowed like sin

Dwight Lee Bates

Out to the line camp at Teepee Ring
Went Jim and I to ride fence till spring

The nights were long, the days passed slow
And Jim began to talk of the Bow
I could tell by that and other sign
That he was hearing the call of the wine

We rolled out one morning, twas cold and bright
And Jim allowed he would go to town and stay oer night
He saddled up his black horse Joe
And hit the trail for Medicine Bow

Along in the night it began to blow

And soon the air was filled with drifting snow

Blast after blast came swooping along

And the wind kept howling its dismal song

The second morning dawned calm and clear

And I kept watching the trail for Jim to appear

And when by noon he did not show

I saddled up and pulled for the Bow

Twas mighty hard going the drifts belly deep

No sign of a trail for the horse to keep

And where the trail joins the road for the stage

I found Jim's horse, reins caught on a sage

Dwight Lee Bates

And as my gaze swept oer the broad field of white

I knew that Jim had become lost in the night

Then I rode round in circles and covered the ground

Until at last poor Jim's body I found

As I sadly looked on his cold white face

I fancied I could see of his old smile a trace

An empty bottle he held in an icy clutch

Lying there dead still in youth it was too much

And as I turned away my heart filled with pain

I swore to never touch liquor again

For an empty bottle, stranger told the tale

Of a true friend and pal lost on the trail

It was just another tragedy of this life we live

Just another case of weakness and the price we give

And as I live through the years and grow old

I will never forget the tale that empty bottle told

Jay L. Johnson (my grandfather)

Somewhere in France

Author Unknown

In Possession of Jay L. Johnson (my grandfather)

In a quaint old English homestead

Down in Devonshire sun kissed hills

Sits a mother grieving for her son departed

To the battlefield that maims and kills

She is thinking of the last embrace and loving glance

Of her boy who is fighting Somewhere in France

Somewhere in France

In fancy she pictures the deadly struggle in the trench

The wounded, dead and dying, the daring and the brave ??

The sons of old Britain falling beside the allied French

Upholding their country's honor, though their reward may be the grave

And as the picture fades she prays God for the chance

To see again see her boy who is fighting Somewhere in France

Somewhere in France

Mercy to the Magdalene

Author Unknown

In the Possession of Jay L. Johnson (my grandfather)

While walking along the city's busy thoroughfares

We often meet the painted women with her wiles and snares

That belongs to the guild of Magdalene of old

And is bartering her virtue for silver and gold

The smiling face that she wears may be but a mask

That covers the anguish and shame of her pitiful task

Her soul may be longing to flee from the body so frail

That to the eyes of the world is labeled for sale

Like the pitcher that too often went to the well

Too boldly she ventured and at last she fell

Like the broken vessel of the Potter's clay

She is used and marred and cast away

Dwight Lee Bates

Why will we the sins of man condone

While a fallen women for her sins can never atone

How rarely is a helping hand stretched forth to save her

Seldom indeed does she receive fair charity's favor

Though now she is but a bit of wreckage on the sea of forever

Perchance she was once a loving mother and a happy wife

Drifting aimlessly along with the human tide

The respect and love that once were hers is now denied

Nevermore can she command the homage due

That pearl beyond price, a woman true

So do not condemn or judge her unheard

or pass her in scorn with a hard bitter word

Just stop and consider that your son or mine

May be the rogue that is buying her wine

Dwight Lee Bates

Ma's Lineage

In August of 1993, my mother or ma as I called her, shown in the following photos, sent me her lineage. This was a month before she died. She wanted me to keep it to prove her descending from the Springers in Europe. They had a Coat of Arms as shown by the knight in the following page from a book piece ma sent. Ma joined the Daughters of the American Revolution to preserve our lineage. My dad had a plaque showing his Coat of Arms before he died. I think he was from noble background also but he never talked about it. Ma always said her lineage was better than my dad's but I never put any credence in that. I knew the relatives came over from Europe to homestead like grandpa Johnson. This is described in another story in my memoirs. My brother Jay has gone on line to research the lineage on both sides of the family. My cousin Buster's wife was putting together a book on the Johnson side. But when I asked about it, I got no answer. I think it has fallen by the wayside. I gave her my info at Pete's Family Reunion but I do not know what happened to the Johnson Book Project. I am glad to see my brother Jay research it. He usually gets to the bottom of things and likes to do research on his computer. I like to do it also. I think when you get older, you want to preserve the history of the family. This is why I am writing my memoirs which is up to 6 books and 2400 pages so far. I can not tell what goes beyond the Springer note shown in the following pages. In the note, ma says: " See Springer Papers and Land Grant from Lord Baltimore and Old Swedes Church Records and Canadian Records and Old Dutch Church Records to find complete genealogy." This is a roadmap to pursue it further. I think ma in a nut shell wanted to show the path to our lineage on her side. I am more proud of grandpa Johnson's homesteading than the knight in shining armor. The things the immigrants from Europe like my grandpa went through to become successful makes me proud. Another story I wrote in my memoirs tells about my descending from this to build airplanes. You can descend from nobility but it is what you did in your lifetime that counts. I know I came from good stock. The following photo I took shows photos from ma's funeral on display.

[Handwritten note: See Springer Papers & Land Grant from Lord Baltimore, & Old Swedes Church Records & Canadian Records & Old Dutch 117 Church Records & etc etc in Complete Genealogy]

DESCENDANTS
OF
CHRISTOPH LOUIS I.,
COUNT SPRINGER OF WALDENBURG,

Son of Louis II., ("the Springer"), Landgrave of Thuringia.

Having given the derivation of Louis II., "the Springer," a brief out-line of his history, together with a short genealogical history of the descendants of his younger son, Louis III., who succeeded his father as Landgrave of Thuringia, we will now take up the line descended from his son, CHRISTOPH LOUIS I., who was named *Count Springer*, and presented with Waldenburg, in Silesia, by the Emperor Henry IV. He was knighted by that emperor in 1092, when but one year old. Knighthood* in the olden

---

*In the ancient German communities there was a distinction between freemen and nobles. The possession of greater estates, or the grants of greater fiefs, had raised up out these nobles "dynasties" or princely families, who formed the high nobility. Many noble families, whose estates were not great enough to sustain this high rank, voluntarily gave up their standing among the inferior nobility, in order to accept that of dependents of the princes. They were often charged by the high noblemen or clergy with special offices, and were repaid for their services by fiefs, which soon became hereditary. Dependents who could afford to render military service on horseback were much more respected than the poor freemen who had not the means to do this, and during the times of the crusades, they, together with the lower nobility, gradually grew into a distinct class known as the equestrian order, or

THE KNIGHT OF WALDENBURG.

Dwight Lee Bates

Bungee Jumping

In 1992, I drove Jay and Julie to Federal Way, Washington to go to the Wild Waves Water Park. We had a day left over from our backpacking trip to Montana which got snowed out. All day we went down a bunch of huge water slides. As we were leaving, we saw a 150 foot crane which you bungee jumped from. Jay said he would pay for one half of the $50 fee if I did it. I readily agreed. They attached the bungee cord to my ankles with straps which I checked carefully. Then I looked down 150 feet which made it hard to jump. They warned you not to grab the bungee cord on your way back up or you would burn your hands. The ground came up unbelievably fast when I jumped. Jay took my picture in mid air. I rebounded 6 times before it dampened out. During the first bounce, every vertebrae in my back popped. After I stopped bouncing, they lowered me to the ground. It was quite a thrill like my sky diving or climbing mountains solo. I would do it again if someone would pay for it. Rarely does the bungee cord break. It is pretty safe. I had the following photo Jay took put on my coffee cup for work. My boss the Chief Engineer saw that and said I was nuts to bungee jump. I see people on TV bungee jump into a canyon or off a bridge. One guy is doing it off an arch in Utah. A lot of people chicken out so they have to push them off. Some people look down and get scared so they do not do it. It is like anything you have to commit to it and do it. I think jumping is better than being called a chicken. Besides if I had not done it, I would not be writing this story for my memoirs now.

Dwight Lee Bates

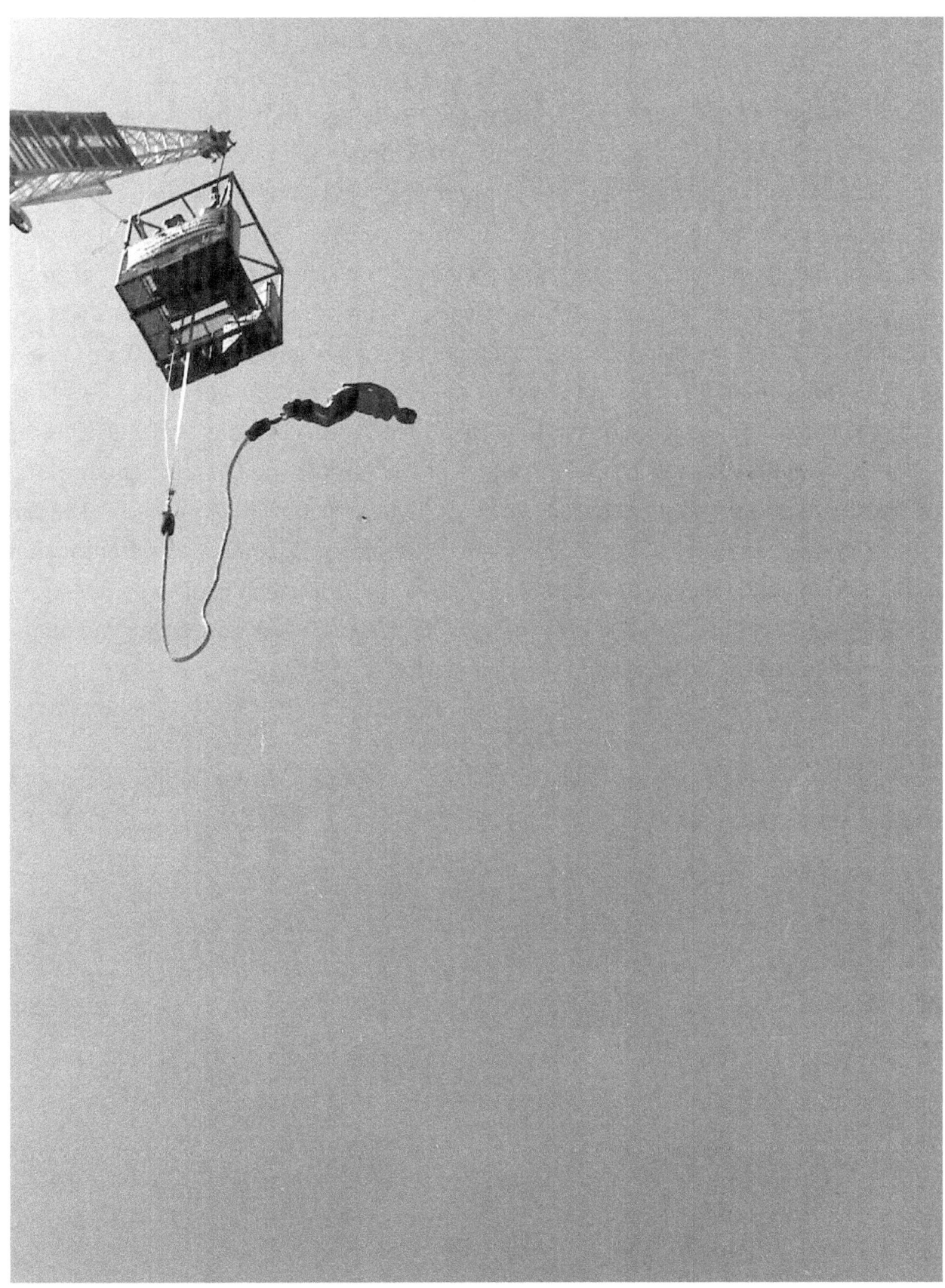

Dad's 80th

In 1995, I flew down to Jay's for dad's 80th birthday. We bought him a belt buckle inscribed Eds 80th on it for his birthday. Then we all went out to dinner at a fancy place Jay picked out. The next day Julie took me to the stable where she rents a horse. My flash going off, when I took the photo of the horse, almost caused a stampede. We played golf the next day. Dad was still playing well. The starter said we were too slow even though we let people play through. I complained in a letter to the mayor and she refunded me everyone's money for the green fees. Dad was happy to see us. I hope I get a party when I turn 80. It was a good break from work at Boeing to go down to San Francisco for the weekend. I flew back on Sunday to go back to work at Boeing. Dad lived 5 more years and died in 1999 at the age of 85. We flew out to Arkansas to see him before he died. Scott and I barely got out of Oklahoma City before they shut down the airport due to tornados. It was called: " The Day of the Tornados." There were about 150 of them. Scott had to fly to about 5 different cities to escape them and get back to Montana. I flew home OK through St Louis on a Boeing 757. The following photos show dad at the age of 80. It was the last time we all had our photo taken together as shown in the following photo.

Dwight Lee Bates

Tomahawk

In 1999, when Diane and I stayed in Ellensburg to see if we wanted to retire there, we went horseback riding at the Tomahawk Ranch. Diane played with a colt she liked. Then they had Diane ride around in the corral to see if she could handle a horse. The woman I know Denise gave me her husbands' favorite saddle. He was an Indian I called Chief. When he saw me using his saddle, he was not too happy. We rode up the pretty Green Valley for about 5 miles as shown in the following photo. At no time did we gallop the horses. I used to ride horses at the A-1 Ranch my Uncle Robert owned. I hayed there during the summer. We used to rent horses at Cheyenne also. One time I went with Dwayne Kremer to rent horses. I rode up behind his horse and swatted it on the behind with my reins. His horse took off running before Dwayne finally stopped it. I like riding horses. Diane liked it too. It is a good way to get in the outdoors. And it does not cost that much to rent one. However owning a horse in Ellensburg costs about $15K a year. A lot of people own them and belong to riding clubs. They ride all over up in the mountains. I think I would rather own a motorcycle. It is easier changing spark plugs than feeding hay.

Dwight Lee Bates

Diane's College Degree

In 1991, my wife Diane graduated from Bellevue Community College with an Associate Degree in Teaching as shown in the following photo. I thought that Diane should get out of being a Dental Assistant since she did not like doing it anymore. If you do not want to go to work, then change your job until you want to go to work. I have always reached for the brass ring. This means go for challenging work. The following is Diane's decision to change her profession:

"I worked for 25 years as Dental Assistant for 3 different dentists. Then I decided to change professions. So in 1989, I went to Bellevue Community College. I started the first year taking Interior Design. I liked the classes but not the people in it. So I changed my major to Early Childhood Education or Special Education. I did an internship with Washington and Bellevue School Districts. I graduated in June 14, 1991, as shown in the following photo. My mother, brother and my husband Lee came to the graduation but my sister did not. I graduated with honors and got a job with the Bellevue School District. I worked on this job for 10 years. I made more money as a Dental Assistant but enjoyed this job better. The reason was that I liked teaching the children from ages 3 to 5. My specialty was children with behavioral problems. For example, one little black boy named James was a challenge. I solved this by taking him outside and having him run around until he calmed down. One mother met me every morning so we could talk about her son. My High School education was West Seattle High School from which I graduated in 1961. I worked my Senior year at Sears. Also I worked an internship at the University of Washington Dental School and the Veteran's Hospital. I also worked at Sun Valley as a waitress with my friend Linda Frankie. We had 9 months of wild times and skied every day. I went to Edison Technical School for 1 year to become a Dental Assistant."

Dwight Lee Bates

Dwight Lee Bates

THE OUTDOORS

Lee's Search and Rescue Help

The Brothers Mountain

When I climbed the South Peak of the Brothers Mountain in 1990, I saw an empty tent with food hanging from a tree at base camp. No one came back to the tent. I opened it up and a cold chill raced down my back. I knew the person was dead. I looked for a body when I climbed the peak but found none. The next night no one came back to the empty tent. At the last switch back before the trail head, I heard a 2 way radio which I knew was Search and Rescue. They asked me if I saw a hiker and I said he is a climber now and I think he fell off the top. Then guy's father started crying. The Search and Rescue looked for a body when they also climbed the peak but saw none. Then they called in a helicopter from Fort Lewis which took them around the bottom of the peak where they found the body. He had fallen off the top 700 feet to his death. It was over hung so we could not see the body from the top.

Lake Serene

My second Search and Rescue help in 1995 was when 3 men went missing near Lake Serene in the Winter while hiking. I called one of the fathers who worked at Boeing and said how they might have fallen off the cliff by the old cabin. I said I almost fell off there. They searched there but did not find them. A week later it warmed up and they found part of a tent sticking up. The three men in the tent were suffocated in their sleep by an avalanche.

Lake Ingalls

The third search and Rescue attempt in 2010 I helped with was a lost hiker near Ingalls Lake. She was lost 2 days in a snow storm with snow up to her waist. Called Search and Rescue to help since I have hiked the area many times. I said she is probably at Stuart Pass since it is easy to miss the trail when it cuts up the cliff to the left by Ingalls Lake. The weather cleared for an hour and the helicopter found her in a grove of trees where I said she would be. The Sheriff's Office thanked me for my help.

Dwight Lee Bates

Solo Climbing Three Fingers Mountain

I was drawn to Three Fingers Mountain because I saw it every day from the Boeing Everett Factory where I built 747s. In August 1988, I drove on a logging road for 30 miles and parked my car at the trail head for Three Fingers Mountain in the Washington Cascade Mountains. I hiked for 5 miles until I came to a wooden shelter where I spent the night. When I woke up in the morning, a curious deer walked up to the shelter within 10 feet of me. It acted like it had never seen a human before. Then I hiked up the trail for 2 more miles to Goat Flats. There I found an old wooden shelter used during World War 2 to watch the Straits of Juan de Fuca for enemy ships. I hiked up to a snow field where I saw Avalanche Lilies in the melt water. I spent the night sleeping out in the open with out a tent. During the night, a climber woke me up in the dark asking where we were. I said Goat Flats. The next morning, I left my sleeping bag and hiked to Camp Saddle where the real climb began. As I climbed up to the glacier, I saw a Pika in the rock scree. Now I could see for hundreds of miles in all directions. Next I climbed along the glacier on the edge of a steep drop off of 500 feet to the foot of the South Peak of Three Fingers Mountain. I had seen this area from 30 miles away at Boeing so I knew where I was. The snow was steep and slushy so it was good to get on solid rock. I found a 6 inch wide ledge leading to the top. The exposure was extreme with a 500 foot drop to the glacier below. I left my pack at the foot of the peak on the glacier. I inched along for 20 feet until I came to a chimney which I climbed to the top at 6,870 feet. The view was spectacular in all directions. I could see Whitehorse Mountain and Mount Pilchuck where I climbed before and Mount Baker in the distance. Also I could look straight down the East face for 2,000 feet. This was the biggest exposure I had seen. I climbed down to my pack and ate lunch. Then I hiked down 4,000 feet and 9 miles to my car at the trail head. I then drove home. I spent 3 days and 2 nights in the adventure. It was really fun and I will always remember it.

Dwight Lee Bates

Lee's Jump

On Friday June 4, 1993, I read in the Boeing News that sky diving was available through the Boeing Sky Diving Club. The cost for a tandem jump was $159 with the Boeing discount. I had wanted to do this for 20 years so I called Rich Branson of the Boeing Sky Diving Club to schedule me for Sunday June 6, 1993.

Sunday I sat around at the Sky Diving Center at Harvey Field near Snohomish, Washington waiting for the weather to clear. I heard horror stories like how five sky divers landed in the Snohomish River and only one lived to tell about it. The jumpers had only one chance and that was to get out of the harness ten feet above the river otherwise the chute acted like an anchor and would drown them. Other lovely stories included running into barns, parked airplanes and power lines. Now I just wanted to get it over with. I watched the chute packer pack a chute. It is very similar to packing a spinnaker on a sailboat. I told the packer how I packed my gloves into my sailboat spinnaker and they went flying into the water when we launched it. The packer said don't worry if the chute doesn't open, just bring it back and he would give me another one.

Tuesday, June 8, 1993, the clouds finally broke. I headed for Harvey Field where I met my jump master Guy. He gave me a twenty minute course on arching my back, and jumping out of the plane. He said the first time you put your foot on the step above the wheel the wind would blow it off. He said don't worry about it just force it down on the step. I thought all the time I am doing this I am hanging half way out of the plane. He said when we get to 8,000 feet, he will lock the door up, yell to me to jump over the step above the wheel. He said don't let me forget to hook you up to me at the four attach points or I would be dead. I said I would try not to forget because I would look pretty silly augering in.

We took off climbing to 8.000 feet and Guy says isn't this fun. I was wondering at the time how I got myself into this. As if he delighted in horror stories, Guy said you know one year ago today my chute malfunctioned and I landed in the Snohomish River. I realized I how Guy could not steer has reserve away from the river. I asked him how he survived and he said he had practiced it in a swimming pool and the trick was to judge perfectly how high you were to get out of the harness. I fussed with my goggles and glasses to get my mind off the river which was growing in width by the minute. Guy gave me his gloves since my hands were cold even though I should have been sweating out of fear. Guy showed me his altimeter on his wrist and said we are at 8,000 feet so let's get ready. I thought why did this happen so soon? I undid my seat belt and crawled up next to the pilot. I looked out the window. What a mistake. Eight thousand feet is a long way up. Guy connected me to the four points which I checked. He locked the window up on the wing. Show time! He told the pilot to turn to the left a bit which I was hoping to keep us from hitting the river. I was talking to myself: "arch, arch and don't screw up." I thought it will be over in five minutes. Guy said stick your foot on the step above the wheel and it promptly blew off as I expected. Next time I planted it good. Even though the pilot had throttled back, I was aware of the noise and fury of the wind. Guy was saying something but I could not hear him. He hit me on the back so I let go and jumped over my right knee out into space. I told myself to arch. I looked up and saw the cool picture of the plane getting smaller and smaller. This was beautiful. My arch rolled us over on our stomachs face down and the noise was not too bad due to the rubber helmet. I looked down and could not see much but green fields. Guy yelled for me to pull me arms in a bit. This did not change much in my opinion. You had no realization of speed and felt like you were floating since unlike bungee jumping (another story) you are not near the ground. I was thinking this was fun when the chute opened with a loud noise. Then complete silence. We had fallen 4,000 feet in 30 seconds at 120 miles per hour. We were at 4,000 feet and the view was gorgeous. I could see the now huge Snohomish River and the landing zone at Harvey field between my legs. I wondered how we would hit that tiny dot? The answer was we would have to maneuver the chute. Guy gave me two straps to pull on to steer the chute. If I wanted to go left, I pulled the left strap. This pulled the back of the chute down and acted like a brake. As a result, you pivoted around the left side of the chute and turned left. We practiced flaring which we would do for our landing. You pull both straps at the same time and it pulls the entire back of the chute down stopping your forward movement through the air. I practiced flaring. As about four minutes had passed, we were over the hangars at Harvey Field. We circled to the left and seemed to be coming in awful fast. Guy yelled: " flare". I pulled on the straps for all I was worth and we did not slow down much. I put my feet down (landing gear) like Guy told me to do. The ground came up fast and I skidded eight feet on the grass with my tennis shoes. When I stopped skidding, I was leaning back so I fell over backwards onto Guy. The parachute lines and chute settled down over us. It took Guy ten minutes to sort us out. Guy said the chute packer would kill him. So I smart assed it by saying the packer told me to bring it back if it did not work. Yuk, Yuk. Guy said congratulations and shook my hand. He asked for his gloves back and I said I probably left them on the plane as I did not remember taking them off. I was losing points with Guy fast. I said my landing was good since any landing we could walk away from was a good landing. I do not think he liked my jokes. I checked my ass and it was still attached. But I had a large grass stain on my pants that I could not hide from my wife. So I decided I would have to tell her that I jumped out of a perfectly good airplane.

My conclusion was bungee jumping was more bang for the buck. Since you jump close to the ground. You fall head first and see the ground coming up fast. Also the fun part of bungee jumping was bouncing up and down six times. My crash landing in the parachute was similar to the landing we had in a hot air balloon when the wind came up to forty miles an hour. I enjoyed steering the parachute into a landing the most. I would recommend tandem parachute jumping over bungee jumping since I thought it was a lot easier to jump when you can not see the ground. For the timid, I would recommend a hot air balloon flight provided the wind does not come up. Also the tandem jump is only a twenty minutes training session rather than the four hours for a static line solo jump. And the guy strapped to your back is not likely to make a mistake since he would die along with you.

Kayaks and Sailing Dinghies

I bought an inflatable kayak after kayaking in a similar one down the Klamath River in Northern California with my brothers and dad. I was the only one in our adventure tour of 30 people to go through 50 major rapids in 30 miles without capsizing. I took this kayak fishing for Trout in Western and Eastern Washington and Eastern Oregon lakes. It would inflate in 20 minutes and did not take up much space deflated. I only paid $80 for it at Fred Meyer. I used it to kayak up and down Portage Bay near our house. I was in the Seattle Times Newspaper when my photo shown following was taken kayaking near the University Bridge on Portage Bay in Seattle where we lived. The inflatable tubes on each side made it comfortable and hard to turn over. Diane and I rented kayaks on Lake Union in Seattle for evening kayak trips. I went on a trip with Scott in his canoe to the end of Lake Kitna near Glacier Park, Montana. I also used to sail and row my El Toro dinghy in the Montlake Cut on Lake Washington. I made a hand trailer for wheeling my El Toro dinghy down to Portage Bay to launch. I kept it suspended from my carport in Montlake. It is now suspended from my garage in Ellensburg. I have sailed it on local ponds. I have taught my friends to sail in my El Toro dinghy when I kept it at Leschi on Lake Washington. I painted the inside of this El Toro sailboat/dinghy with epoxy paint similar to the paint I used on my airplane I built. I paid $100 for it when I answered an ad on Lake Union at a house boat. The girl had just kicked out her boy friend who owned it and was selling it to get even. She asked what it was worth. I said $100 when it was worth a $1000. I quickly hauled it out of there before the boy friend came back. Also I had a Nelson sailing dinghy in Port Orchard I used to sail all over Sinclair Inlet where our boat the Catnip was moored. In the San Juan Islands, I used this dinghy to sail around our anchorages and to go a shore. I took an umbrella to use as a sail when using it without the sail up.

Dwight Lee Bates

Dwight Lee Bates

Dwight Lee Bates

Dwight Lee Bates

Dwight Lee Bates

Rampart Ridge

In 1968, Diane and I hiked to Rampart Ridge near Snoqualmie Pass. We took a heavy Boy Scout pup tent that her dad gave us. It was rubberized fabric so it was hot inside. We had great fun hiking around the individual tarns in the area. There was not many bugs to Diane's relief. No one else was there. A local trail guide describes the hike as the following: "There were some blueberries left near Rampart Lakes, although they've been well picked by previous hikers. The mile or two of trail leading up to Rachel Lake is steep, rough in parts. Rachel Lake is really pretty. Look for piles of branches that trail crews have placed to direct hikers to the main trail. The lower part of the trail has several pleasant spots overlooking waterfalls and smooth granite. This is a lot of fun for a photographer with a tripod. A hiker was once heard to say arriving at Rampart Ridge :"If this is not heaven when my time comes I will refuse to go." At about 2 ½ miles, the trail runs into the valley head wall and climbs 1,400 feet in a mile which is steep. However, at 3 ½ miles you reach beautiful Rachel Lake at 4,700 foot altitude as shown in the following photo. You can camp any where. Turn right at the lake shore and gain 500 feet to a saddle at 5,200 feet. Turn left and at another mile arrive at 5,100 feet and Rampart Lakes. Tarns or ponds are every where. There are various sizes and depths. You can swim out to islands. Wander through meadows, around rock buttresses into nooks and crannies. Climb to the top of Rampart Ridge for a spectacular view. You can see: into Gold Creek, Snoqualmie Pass Peaks, Mount Stuart and South to Mount Rainier." The total round trip is 10 miles and the elevation gain is 2,400 feet.

Dwight Lee Bates

Long Draw

On the farm in 1956, a favorite place to go to fish for the day was Long Draw. We approached it from the Cache La Poudre River side but a better road is probably through Estes Park. The following photo from Wikipedia of the Cache La Poudre River near Long Draw is an area we fished often. When we did it, there was no official campground. It had fire pits of rock circles where people made fires. The fishing was wild Trout also. I remember dad was fishing below a falls in a deep hole. He said he had hooked about a 5 pound Trout but could not catch him. I watched as he would hook it and it broke off several times. At least we got to see how big he was. I fished the creek below the falls. I hooked a 12 inch Trout but he broke off. Then I hooked him again and landed him. My original hook was still in his mouth. Why these fish kept biting probably meant that they were hungry and wild. We never camped there over night. But I would not want to camp there using dad's methods. He would build a big fire. Then he would lay 2 quilts on the ground and the whole family in mass slept between them on the hard ground. This was awfully cold. He must have thought that our body heat would keep us warm. I remember dad's awful method of starting a fire in a rain storm. He would get a 5 gallon can full of gas and keep throwing it on the fire until it got going. We never had sleeping bags and air mattresses. We never had any money to buy them. We got our first sleeping bags in Cheyenne when dad had a steady job. But I would not trade those days for anything. Dad was an excellent fisherman with flies. He would catch a lot of wild Trout. In those days, all we had were wild Trout. Those were the good old days before our country got so populated that we rely on hatchery fish now a days. When we went fishing, we would not see anyone else. It was wild country. I miss it.

Dwight Lee Bates

Rainbow and Therriault Lake

In August of 1998, we made a trip to Scott's to go to Rainbow and Therriault Lakes. They were located in the northwest corner of Montana near the Canadian border. We made our base camp at Lake Therriault and went canoeing there. We took our vehicles as close as we could to Rainbow Lake. We had to walk about 4 miles before we could see Rainbow Lake. But the worst was yet to come. We had to go down a narrow rock strewn gully as shown in the picture. I made as much noise on the trail to discourage Grizzly Bears. The dog kept herding us down the trail like were sheep. It was a sheep dog. We did not catch any fish at Rainbow Lake so we hiked out. About one half way up, the boys above me kicked loose huge rocks. They just missed me by a foot as I jumped out of the way. It was a 500 foot climb. We canoed on Lake Therriault but did not catch any fish. However the lake was beautiful. At camp, some idiot fired rifles near the campground. Where is a forest ranger when you need one? The country was pretty but we did not catch any fish. Scott had caught some at Rainbow Lake before. I was nervous being in Grizzly country since I had my run in with a blonde one at Diablo Lake campground in Washington. I am too tough to eat though. A grizzly probably would not want me. I was ready to tie someone's shoe laces together on the trail so I could out run them. You do not have to out run the Grizzly only the poor guy with his shoelaces tied together.

Dwight Lee Bates

Lake Hamby

In 1998, we went to Lake Hamby, Montana backpacking. It is on the border between Montana and Idaho. Jasin, Emily, Scott, Jay, Brett, Jon and I, shown in the following photo, drove to the Big Hole Battle Field. This was where Chief Joseph defeated the US Calvary that was trying to chase him and his tribe down. They wanted to force them back to their Indian Reservation. This is the war that Chief Joseph said after being defeated in Montana: "I will fight no more forever." I have been along most of the route the Indians took. We drove on through Wisdom and tried to cross a river. I said I would try it with my Cougar car. After getting to the other side, we decided it was too deep. So we drove 20 miles along a primitive Jeep road. At the end of the road, a Jeep Club applauded me for getting my Cougar car up the road. They said they never saw a car make it all the way. We forded a creek and went straight up. Brett and I lagged behind like we always did. About sunset, Brett and I reached the top. We hurried down the other side to find the rest of the party making camp. The next morning, we hiked about 5 miles to Lake Hamby. Brett got stuck in a mud hole. I had to pull him out and his boot came off. When we finally got to the lake, the rest of the party was making camp. The mosquitoes were awful. We tried to hike to a higher lake to avoid them. I watched as a gust of wind blew them away. Back at Lake Hamby, 2 guys showed up on 4 wheelers. They brought in a rubber raft which they took out in the middle. They caught more fish than we did. The next morning Jon and I got up at sunrise to go fishing. I caught a 13 inch Brookie. Back at camp people complained that they had slept on a bog. Then we decided we had too much food to pack out. Jon hit a can of cheese with his hatchet. It took off flying through the air due to compressed air. We hiked all the way out and drove to Jackson. When I arrived, everyone was sitting outside the General Store on old wooden chairs just like in the Old Wild West. Then we drove to Missoula where we all piled into one motel room. We went out to dinner that night and I gambled. I won $5 so I quit while I was ahead. The next morning, we split up and drove home.

Dwight Lee Bates

Climbing the Brothers

In August of 1992, I climbed the Brothers South Peak near Hood Canal in the Olympic Mountains in Washington. Since the beautiful Brothers Mountain is visible from Seattle, I always wanted to climb it. The South Peak at 6,866 feet is the highest of the 2 peaks. A girl I worked with said her husband would fly me over the peak in the Boeing Flying Club's Cessna 172 to scout the route. We flew the peak in about an hours time. Her husband said: "You are not going to climb that by yourself are you? You are crazy!" I was able to see the route was a couloir with a pinnacle on the left as a landmark. The couloir which is a gully still had snow in it which concerned me as I did not want to pack crampons all the way up there. I paid her husband $60 for the gas for flying me over the route. I figured it would be a one day climb up and back from base camp which I also saw from the air as shown in the following photo. The girl I worked with at Boeing said:" Lee you are going to kill yourself. Why don't you hire a helicopter to land you on the top so I do not have to worry about you?" I said: " I like the experience of climbing the mountains alone. I can not climb the big mountains because I can not breathe at high altitudes due to the asbestos in my chest from the shipyard. Besides I have climbed the following mountains the same height solo: Three Fingers and Pugh. If I make a mistake, I probably will not survive. Also when I am on top, it is like a religious experience. I feel next to God. That is the way I want it." I am sure she did worry about me. In August, I was at Mount Rainier hiking up to the Muir Snowfield when the weather improved. I drove back to Seattle to get my gear and drove to the trailhead. I saw only one other car there so I was happy that I would be alone. I passed Lena Lake shown in the following photo on the way to Camp Brothers. This is a pretty lake used by day hikers. When I got to Camp Brothers, it was dark. I ate dinner and laid out on the ground without a tent. I was awakened twice at midnight and 3 AM by an unbelievable loud wailing noise 50 feet away in the brush. I later concluded it was a Sasquatch (see other Sasquatch stories in my memoirs). For those people who disbelieve that there is a Sasquatch, I can only hope that you are laying out in the open in the darkest of nights like this and hear this really loud wailing noise. This would make you a believer. I did not know what the noise was and mentioned it to a guy at work. He said Sasquatches make a noise like that and my wife is reading on book on it. I called her up to get the title. I went and bought the book. Everything fit for it being a Sasquatch. Then 2 years later, I was watching TV. A Lummi Indian Police Officer recorded the noise of a Sasquatch he saw. He played it back on TV. It was the exact same noise that I had heard. I called him up and he described it. It had no neck, was huge and hairy. This confirmed that it was a Sasquatch. He said he hated to tell the story since people refused to believe him. I believe him. Earlier I noticed an empty tent and food hung up in the tree. I yelled: "Is anybody inside?" No one responded so I put my head in. A cold shiver ran down my neck and I knew the climber did not return because he fell off the top. The next morning I headed for the top. My landmark was the obvious pinnacle to the right. I had an altimeter with me so I logged the altitude when I took the photos. One photo shows Hood Canal from 5,100 feet. The going is getting tough because of the loose rock. It is a good place to break an ankle. I cached my pack here to lighten my load to the top. It will be all sweat from here. At 6,300 feet, I was at the pinnacle landmark I saw from the air. I am right across from it. The loose rock was dangerous. I saw a gendarme which is French for a high rock pinnacle. At 6,500 feet in the couloir, I could see Hood Canal in the distance. One of the following photos was taken at the top of the couloir at 6,600 feet. It shows how steep it is. I could see the Great Bend in Hood Canal and Vashon Island from the summit. After I left the couloir, I had to climb up solid rock to the summit. I looked over the edge looking for the body of the missing climber from the tent at base camp. I saw nothing. The climb to the summit was tricky so I marked the foot and hand holds with surveyors tape. This would enable me to down climb and see where the hand and foot holds I used to climb up with. This made it a lot safer. I figured the missing climber probably did not do this. Maybe he was a beginner? One of the photos shows the North Peak of the Brothers at 6,800 feet which is visible from Seattle. You can climb over to it using a traverse but I figured it would take too long and you might need a rope. Besides the South Peak is 66 feet higher and is the official summit of the mountain. I put my camera on a rock and took the self portrait of me on the top at 6,866 feet. You can see the Mildred Lakes area, which is another story in my memoirs, over my left shoulder. You can also see Mount Cruiser and

Mount Alpha which we climbed as talked about in my other story. I climbed back down slowly using and picking up the surveyor tape I marked my route with. I carefully slid and climbed down to my pack where I drank my fill of water and ate. It was mid afternoon. I still kept looking for the missing climber's body. I saw nothing so I figured he had to have fallen off the top. If he did this, he was dead since it was a 700 foot fall. You can see this face of the mountain from Seattle. I got back to Camp Brothers about dark. I filled my canteen and purified the water. I checked the tent in the dark and no missing climber was in it. The food was still hanging up in the tree. Now I knew he was dead. I ate dinner and laid out on the trail. I figured the Sasquatch would pay me a visit again but he did not. I figured he could have easily killed me as I slept out in the open. But he did not so I thought he was just trying to scare me out of his territory. The next day as I got to the last bend above the trailhead, I heard a 2 way radio. I knew it was Search and Rescue looking for the missing climber. They said: " Did you see a missing hiker? " I said: " He is not a hiker and he is a climber because I saw his empty tent at Camp Bothers Base Camp. His food was still hung up in the tree. He did not come back the next night either. I figure he is dead from falling 700 feet off the top. I looked for his body all the way to the top and back down but saw nothing." The guy next to me started crying. They said he was the boy's father and his boy had never climbed before. I went on the Link Lake backpacking trip with my relatives in Montana. This is another story in my memoirs. I called the sheriff's office from Montana. I said: "Did you find the missing climber on the Brothers Mountain?" They said: "Are you related to him?" They said they could not tell me what happened since I was not a next of kin but there was an article about it in the Bremerton Sun Newspaper. So when I got back to Seattle from the Link Lake trip, I went to the library to look at the Bremerton Sun. The article said that Search and Rescue climbed the Brothers Mountain looking for him but found nothing. They climbed down to the base and called in a helicopter from Fort Lewis which picked them up and flew them to the base of the mountain. There they found his body. He had fallen 700 feet to his death from the top like I thought. The mountains face is overhung which is why we could not see his body from the top. I figured he did not mark his holds on the way up like I did. As I was developing these photos, the girl at the camera shop told me about this 33 year old boy who was engaged to be married. This girl knew the girl he was going to marry. He had never climbed a mountain before. I am glad that I helped

Search and Rescue find his body. That way his loved ones had closure. When I got back to Boeing, the girl whose husband flew me to scout the route was glad to see me. At first they thought I was the dead climber. It was all over the news. Then I showed them the photos that are following this story. They said I was nuts to climb alone. They said the dead climber proved that. Then I told them about the Sasquatch visits. Then they really thought that I was nuts. Later I showed them a photo of me bungee jumping. Then they really started to wonder?

Dwight Lee Bates

Dwight Lee Bates

Dwight Lee Bates

Dwight Lee Bates

Dwight Lee Bates

Dwight Lee Bates

Dwight Lee Bates

Mildred Lakes

In 1975, a group of fellow Engineers from Puget Sound Naval Shipyard and I decided to go backpacking at Mildred Lakes. It is on the Olympic Peninsula. We wandered off the trail going in since it suddenly forked. I got us back on track with my compass. The trail or path was terrible. The trail guide said the following: " Three beautiful backcountry lakes tucked in a hidden valley surrounded by craggy summits await at the end of this arduous hike. It is a boot-beaten path that leaves you worn out. The Mildred Lakes Trail was forged by tenacious fishermen and expert hikers. An entanglement of roots, slippery rocks, and an up-and-down course that defies logic. This hike is about the destination, not the journey. But boy is the fishing good. That's what lures most visitors to these aquatic gems. Everything you've heard or read about this hike is true. It is tough. Never officially constructed, this trail resembles little more than a way path, especially the second half. We're talking about winding through recovering wilderness since the area was logged decades ago. The first mile or so is easy enough. But once you cross a large avalanche chute, the misery begins. Around fallen logs, over fallen logs, and straight-up rocky and steep slopes the trail climbs a 3,200 foot ridge. Mount Pershing looms in the southeast. Through hemlock and huckleberry, the trail hightails it off the ridge. You drop 250 feet into a cool forested glen graced by a rushing creek. You cross it on a questionable log or wade it." We should have waded since the bark gave away on the log and one of our guys slipped. He caught himself, but I could not pull him up. So he dropped his pack about 20 feet to the creek below. As I pulled him up, he said he could not breathe. I looked at his chest and it was badly bruised. The wind was knocked out of him. We recovered the pack and everything was OK except it was soaked. Within two minutes, we came to another creek crossing but this one much easier. Now the way got tricky and even tougher. "Ascending a steep ridge, upward mobility is temporarily halted upon coming to a ravine. Using roots as handholds, drop 25 feet into the dank draw before resuming the taxing climb. After clambering over rock, root, and ledge-and gaining 1,000 feet in elevation-the insanity ends on a 4,100 foot heather-draped ridge crest. You wander along the open ledge, enjoying excellent views of Mount Cruiser and the Sawtooth Range. But no lakes are visible yet. Your punishment isn't actually over. You drop 200 feet through marshy mosquito breeding grounds before finally coming to the first lake at an elevation of 3,800 ft. It is set in deep old timber with a backdrop of rugged summits. The scene is serene and it is one of the prettiest areas I have ever hiked to. The mountains were jagged with points on the peaks. You have to scramble over more root and rock to the far side of the lake for good fishing. You share the abundant trout with the resident Osprey." Later after pitching our tents, we checked out the other 2 lakes. To reach Upper Mildred Lake, one of the largest backcountry lakes in the Olympics, we had to cross Lower Mildred's inlet stream and follow a rough path for 0.3 mile. The third Mildred we reached by following a rudimentary path north from the upper lake's outlet creek. After catching all the fish we could eat, we decided to climb Mount Alta. We scrambled to the top and could see the fog closing in on Mount Cruiser. We did not have a rope but the people camped next to us did. They climbed Mount Cruiser and said it was very exposed. As shown in the pictures, it is a real Tooth. We spent 2 days and hiked out. This time we waded the creek. I am glad I did it when I was young and that I carried a compass. Otherwise I might still be there wandering around.

Dwight Lee Bates

Dwight Lee Bates

Link Lake

In 1992, I picked Jay and Julie up at the Sea Tac Airport in Seattle, Washington after working an early shift at Boeing. We drove the 544 miles to Scott's place in Whitefish, Montana in about 10 hours. Since I picked them up at about noon, we got to Scott's at 10 PM. I bought Julie a model horse at Saint Regis, Montana since I heard that she liked horses. At Scott's, we met dad who had flown in. The weather forecast was for rain but we were committed. We fished at Murphy Lake near Scott's place of work at the U S Forest Service. But we did not catch anything. We went back into Whitefish to warm up and eat. We ate at a neat place in downtown Whitefish. The next day we drove to Link Lake to backpack in. On the way, we saw a Cougar run across the road ahead of us. The kids rode with me in the Cougar Camper because just known to us kids the Cougar had treats. It had candy bars, cookies and potato chips. We packed in about 1 ½ miles to Link Lake. I went back from the lake to my car with Brett to get my raft for fishing. We met 2 guys who had just seen a Grizzly Bear run by our camp. Good thing it did not encounter Brett and me   (See Grizzly Bear story in these memoirs). We made Grizzly Trees for the kids to climb up in case we were attacked by a Grizzly Bear. I tied Scott's shoelaces together again. We made the Grizzly Trees by cutting off branches so that the kids had a ladder of branches to climb up. That night as we were sleeping, I heard a noise. I got up and saw that we were in a 100 Year Montana Blizzard in August. So I got up periodically knocking the heavy snow off everybody's tent so that they would not collapse. In the morning, our gear was hard to find since everything, as shown in the following photo, was covered in snow. The kids only had tennis shoes on so they got cold. We could not find the trail out because the snow had covered it. We finally wandered out and everyone was cold. The kids and I warmed up with the Cougar's heater and the sugar from Lee's stash. Then we went to dinner at Whitefish. We then drove back to Seattle where Jay, Julie and I went the next day to the Wild Waves Water Park in Federal Way. There I bungee jumped which is another story.

Dwight Lee Bates

Dwight Lee Bates

<u>Cutthroat</u>

In 1989, we went to Bellas Lake, Idaho with dad, Emily, Brett, Julie, Jason, Scott, Jay and Jon. We had gone to this lake many times fishing, climbing and backpacking. My Memoirs tell of Sasquatch runbys and attempting to climb Pyramid Peak in this area. This time I packed in a small raft and fished out in the middle of the lake. I got a huge strike so I knew there was big fish in it. But this big Trout did not swallow the Power Bait so he got off. Also a guy came in by horseback and chummed by throwing bait out in the lake. I saw him catch some big fish so I knew they were in there. In the middle of the afternoon, I was standing by my pole leaned against a small Pine tree and talking to Scott fishing next to me. I was using Power Bait on a hook with a sliding sinker. The sinker sits on the bottom and the Power Bait floats up so the fish can see and smell it. The fish swallows the Power Bait so they are not getting off. They hook them selves. I showed the kids how to do this and they really started catching fish. All of a sudden, my pole doubled over. I yelled " I got a big one." Everyone came running except for dad who was fly fishing at the end of the lake. I played him for about 20 minutes and he put up a big fight. He kept diving down by a tree under water. He was trying to wrap the line around this tree it to break off. But when he did this, I pulled his head up. After he was tired out in 20 minutes, I told Scott he was ready to bring in. I did not have a net so I asked Scott to help me. As he got near the shore, Scott grabbed the line and pulled him in. I let out big war whoop. He weighed 3 ¾ pounds and was a wild Cutthroat Trout. I had him mounted and he is in my shop on the wall as shown in the following photo. The guy who mounted him did a table mount but I liked the wall mount better. It cost me $200 to have him mounted in Snohomish, Washington. I knew better than have him mounted in Seattle. It would have been very expensive there. I remember when the big Trout was in the creek on a stringer, Jay looked at him and said: "Beautiful." I got to eat the meat after they skinned him at the taxidermist. It was delicious. This fish was really smart. He was missing part of his jaw from fighting with other male Trout during spawning. Also I caught a 3 ½ pound Brown Trout on the A-1 ranch when I worked there for the summer as a kid. I dropped my line in over the bank in Pass Creek into the hole about dark. I was hidden so the fish did not see me. Then something tugged on my line and was swimming away with my worm. I pulled and all hell broke loose. He started thrashing in the creek. I knew he would break off if I played him. So I pulled him up the 5 foot bank. Then my pole broke so I pulled him in by hand. But I never got to eat this one. I kept him in our freezer in Cheyenne showing everyone until he got freezer burnt. Then ma threw him out. I did not have any money to mount him. Ah the big fish! I caught sixty seven 2 ½ pound Lahontan Trout at Mann Lake, Oregon but they did not fight like these big Trout. They fought well but the heavier fish could break your line in a second. See the following photos showing my big fish. These big Trout fought like Steelhead. My biggest Steelhead was 15 pounds caught on the Snake River in December. I played him for 30 minutes to tire him out. Butch the skipper of our guided boat said I played him perfect. Dick my fishing partner that day also caught a 15 pound Steelhead. But since his fish was 2 ounces bigger than mine, I had to buy dinner due to the bet we had. It was a dinner I was happy to buy. This fish is shown in the following photo. The Sea Run Cutthroat I caught in the Skagit River shown in the following photo is also a beautiful Trout. I have hooked big fish a number of times that took off on a run. To avoid losing all my line, I had to break them off. You could tell by they way they pulled that they were huge.

Dwight Lee Bates

Dwight Lee Bates

Mulie

In 1958, I shot a big 4 point Mule Deer buck on Wood Mountain. It took us all day to drag it down. I mounted the horns and made a gun rack out of it. Recently I saw a TV show which showed a guy shooting a Mule Deer buck smaller than mine near Cody, Wyoming. The contrast is remarkable. I borrowed my Uncle Robert's 30-30 rifle and climbed the mountain for my 100 yard shot. The Mulie was leading a herd of about 10 does. The shot in the shoulder was so good that I could see blood when the bullet hit. I only needed one shot. I only shot a big buck one time since the 2 point deer were a lot better eating. My Mulie did not cost me anything but $10 for the hunting license. The guy who recently shot the Mulie was from out of state. He got his permit for a special area by points. They saw the big buck on the first day but took 3 days to finally shoot him. The guy had to pay for a guide and a spotter. He had a rifle made for him which was waiting at the guide's shop. It came with a case. His whole hunt probably cost about $5k. My 30-30 had iron sights. I had no spotter. His shot was from 700 yards with a 4 to 20 power variable scope. The spotter was even telling him the windage which he adjusted on top of the scope. When he first hit the buck, the spotter told him to shoot again. The second shot brought him down. How things have changed. You buy your way into a big buck. I would never think of shooting 700 yards. The only thing I would have need a spotter for was to help me drag the Mulie down the mountain. I was shooting for meat on the table. But I confess that I wanted a big buck. I did not have to shoot him high up on the mountain. I could have shot a big one lower down so we did not have to spend all day dragging him down off the mountain. I am sure my dad and my brother Jay have not forgiven me for the hard work. I think ma threw the horns out. They were not in the house when I left for my job at Boeing in Seattle, Washington. I think she also threw out Jay's fly rod too. The photo following of a Mule Deer is from Wikipedia.

Dwight Lee Bates

Sisters

As you drive South towards Bend, Oregon on a rock hounding or fishing trip as we always do, you see a chain of volcanic mountains. The first is Mount Hood. The next is Mount Jefferson at 10,497 feet. Then you see the Sisters Mountains. The first is the North Sister at 10,085 feet and the next is the Middle Sister at 10,056 feet as shown in the following photos. Then you see Broken Top and the South Sister at 10,358 feet. They are good climbing mountains with a combination of walkups and technical climbs. The photos following with the sunset and view of the Middle and South Sisters is from my nephew Jon who did a technical climb on the North Sister. My nephew Jon also guided on Mount Hood 11,239 feet which is a serious glacier climb. The lodge there called Timberline Lodge is spectacular. It was built by the CCC boys in the depression. The U S government gave these out of work men jobs and kept them in camps. We went to a quilt show in Sisters, Oregon which is a neat small town. Also, you see the beautiful Mount Jefferson shown in the photo. These mountains make this area a good place to live if you like the outdoors. We did a fishing trip near Mount Bachelor one year but did not catch any fish. Bend grows bigger every time I drive through it. My brother Scott lived there one summer. My nephew Jon met his wife Heather there. My cousin Pete had a golf course 70 miles to the east where we frequently played golf. However he had to close the golf course since it is a depressed area with no work so the people do not have jobs. The environmentalists shut down the saw mill and there is little logging.

Dwight Lee Bates

Dwight Lee Bates

Dwight Lee Bates

Dwight Lee Bates

Baca's Boat

In the 90s, my friend John Baca had a power fishing boat in the Everett Marina. We worked together at Boeing. He would say lets go fishing. We would take off after work and fish until after dark. The following photo shows an afternoon we spent out on the water. His favorite spot was Possession Point as I described in other stories in these memoirs. We used to catch 15 pound King Salmon on his down rigger there. Another friend of mine had his boat at the Tulalip Indian Reservation Marina. We would also go there after work. We would stop at the community crab pot and get some crabs. Then we would head up to Camano Island and beach the boat. Then we would dig clams. We had a big pot we would fill with sea water. Then we would build a fire. We put the crabs and clams in our pot and on the fire. We then would eat like kings with fresh sea food. We were under a lot of stress at Boeing and this helped relieve the tension. I often think back on those days now that I am not near salt water. I do not miss the crowds on the west side of Washington but I miss the salt water.

Dwight Lee Bates

Dewey Lake

In 1992 after work, I drove over Chinook Pass to the Natches Peak Trail. I took the following photo of Mount Rainier from the road above White River. Then I took the photo of the creek I saw from the road. The leaves were changing. The Natches Peak Trail starts from the parking lot at the top of Chinook Pass at 5,432 feet. The trail is about 3 miles around Natches Peak at 6,452 feet and did not have much elevation gain. You could look down on beautiful Dewey Lake which is the following photo I took. I saw a Marmot on the trail so I took his photo. The hike stirred my emotions after working long hours at Boeing designing the 777 airplane. I enjoyed the wild beauty and stunning wild settings. Dewey Lake was a big lake that a friend and his wife camped at. Dewey is one of the monarchs of alpine lakes. It is amazing scenery with Natches Peak towering overhead and a rich forest. Dewey Lake is famous for its Trout fishing. It is stocked regularly by plane. The plane dives down over the lake and releases Trout out of a tank. This spot is ideal for a picnic. It is a good place to take your wife because there is little elevation gain. Also there is always plenty of spaces at the parking lot. I have gone on hikes like Ingalls Pass and Cascade Pass where lately the parking lot is full. There is getting to be too many people in Washington. Don't they know that it rains here?

Dwight Lee Bates

Dwight Lee Bates

Dwight Lee Bates

Owyhee River

In May of 2014, I met my brothers on the Owyhee River in Oregon to fish for Brown Trout and rockhound. We got a good camp on the river and fished for 3 days. Scott and I used Mepps spinners and flies and Jay used PMD flies. We all caught fish but Jay must have caught 11 Browns. They averaged about 2 ½ pounds. A thunder storm sent us to our vehicles. The weather was cool and not hot. We saw a bank Beaver swim up and down with willows he was putting in his den to eat later. They eat the bark. We made fires at night from the wood people left behind. Then we went to Succor Creek where we had a great camp. Our camp and the view of the spires from camp is shown in the following photos. We ate mostly hot dogs and beans at this camp. We drove around above Succor Creek to find Thundereggs in a cliff. A rockhounder told Jay about the location. We gave a small boy a couple of Thundereggs hoping that he will be a future rockhounder. Then we drove over the country to the west of Succor Creek looking for Jasper. We then drove down Leslie Gulch. The Succor Creek area was full of thousands of black butterflies. We saw a lot of soaring big birds and deer. We are afraid that this land will be made into a National Monument by Obama. This brings rules keeping rockhounders and fisherman off the land just to return the land back to the animals. For what purpose this happens we do not know.

Dwight Lee Bates

Mount Stuart

In 1995, I decided to climb the south side of Mount Stuart. It was a summer where we did not have a summer. It rained in Washington all summer long. I saw a 2 day window that might work so I left for the mountain. I spent the night at Long's Pass and made that my base camp. From this side, it is a walkup as shown in the last photo. From the other side, it is a technical climb. I got half way up when a storm hit. A gust of wind knocked me off my feet. So I aborted the climb. I always wanted to go back but never have. My nephew Jon and a friend climbed the North West Face route in 2005. The following photos are showing his successful climb. I have seen this route from the air when I flew by the peak before my climb. It is spectacular with hanging glaciers and rock buttresses The route Jon took is a 1,500 rock face between the North Ridge Buttress and Stuart Glacier Couloirs. The first ascent was by David Beckstead and Paul Myhre in 1970. Jon's route is mostly Class 5. It follows a curving pattern on the rock face as shown in the following photo from my Fred Beckey's Cascade Alpine Guide book. It is his Columbia River to Stevens Pass book. Jon climbed the Gendarme Direct Variation Route. The climb is Class 5.9. Jon camped below the face before his climb. One of the party is climbing the rock face with crampons on. This is not recommended. On a mixed rock and ice climb, you are supposed to take the crampons off so they do not slip on the rock. Prusik Peak in the same area in which Jon has also climbed. My brother Jay and I hiked by it as told in another story in these memoirs called:" The Enchantments." I admire Jon for all his climbing which I wanted to do. I love to rock and glacier climb. As I get older, I can only rocking chair climb and wish I was along on Jon's climbs. There is a danger since 6 experienced climbers have recently disappeared on Mount Rainier. They think an avalanche killed them since they hear avalanche transponders going off. But they do not want to go there to find the bodies since the ice is still avalanching down. My 2 brothers and I climbed Mount Rainier in 1980. The next year 11 climbers were killed on our route. This is explained in another story in these memoirs. I do not feel you should give up climbing just because of the dangers. I still love it from my rocking chair. Also I see Mount Stuart everyday out our view in Ellensburg. The mountain wants me to climb it.

Jon's Comments on Mount Stuart Story

Thanks Jon. I will put this story in my Memoirs book since it is meant to inspire a 6th grader to do what we did.

Lee

Wow, that is a really good write up of the mountain, thanks for including me in the stories. Just for clarification, we camped at Ingalls lake, climbed Ingalls Peak's east ridge. Then the next day we traversed all the way around the peak and we actually climbed the entire north ridge of Stuart with the gendarme direct variation (5.9) I believe, I remember the dihedral being pretty hard, especially at altitude. We stayed away from the glaciers because they were releasing huge amounts of ice that were thundering down the glacier to the left of the ridge. Our big mistake was not having crampons on because traversing the lower glacier to the toe of the ridge was really challenging with only an ice axe and hiking boots, but we were trying to fly light. Listening to you and dad and uncle Jay talk about climbing around the campfires growing up was my main motivation to want to climb as much as I have. I would read the books you would circulate to my dad and dream about the adventures to be had. I appreciate the childhood I had and all the time I spent in the outdoors. I hope I can do the same with my daughter. Thanks.

jon

Dwight Lee Bates

Wild Fire Fighting

Fire Boss

In 1963 in the Craig District BLM office, I got to be a Fire Boss on my own fire. The Fire Boss on the fire I was on sent me and my pumper to fight another fire 30 miles away. When I got there, it was a brush fire and a lot of concerned ranchers were there. I called the Fire Boss on the fire I had just left on the radio and said can I hire these ranchers to work for us fire fighting so I can contain the fire? The wind was blowing about 40 mph and I was concerned that it would spread from the 300 acres it had already burned up. The Fire Boss said yes so as Fire Boss, I took all the names down of the ranchers who were there and recruited them to contain the fire. These ranchers had 2 Cats which we used to make a fire line. In our weekly BLM meeting, the Fire Boss commended me for containing the fire. We sent checks to the ranchers for their labor.

Lee

Loss of Pumper

I was as the boss was assigned to fight a fire near Craig, Colorado. I took the pumper and 3 guys with me. When we got to the fire, it was too huge to fight. Since no one was at the office, I told my crew to stay put while I drove back to Craig to get help. When I arrived at the office, I was jumped by the Assistant District Manager. He said: " How did you burn up the pumper?" I said: " I told the guys to stay put and came for help." We then drove out to the fire so he could scope it out. At the fire, the 3 guys were standing by the road. They said they tried to cut the fire off by driving the pumper along a narrow ridge. Then the fire crowned and they could not turn the pumper around on the narrow road. The crowning fire chased them down the hill as they ran for their lives. The pumper was over taken by the fire and burned up. There was an investigation and I told what had happened. The guy in charge of the pumper was not invited back to work the next year at the Craig BLM district.

Lee

Dwight Lee Bates

Wheatland, Wyoming Crown Fire

The biggest problem on the Wheatland fire was a lack of experience. Also we did not have fire shelters. Also the fire leader kept putting us on putting out hot spots instead of making a fire line. I told him before the crown fire incident that he was having us do worthless things that were unlike any fire fighting that I had ever done. We only had one Cat for making a fire line. I told the Cat driver that he should try to make line in front of the fire. Then I saw him right in the flames with our only Cat which he owned. I told him he could lose his Cat by getting into the fire rather than getting out in front of it. None of my 30 man crew knew how to fight a fire. I had to give them on the job training. The only good part of the fire was it was right next to a tavern way out in the middle of no where. My brother Jay probably knows the tavern that I am talking about. We went there for beers at night. Also we were fed well since the National Guard was there with their field kitchen. However, the food was so bad that, a friend of mine and Jay's, Mike was served spaghetti. When they put it on his plate, it was all clumped together. So he grabbed it off his plate and started beating it against the spaghetti pot. All my 30 man crew had red backs from the retardant landing on us. Also some of the guys got on the wrong side of the trees when the B-17 was dropping on us. All 30 men were panicked after running up the hill from the crown fire. One guy tried to drive off in our Jeep before we grabbed the keys away from him. There was people hiding under the jeep from the fire. He would have run over them. Also there was no place to drive to since we were surrounded by fire. I had never seen 30 men panicked before. I hope I never see it again. Everyone turned into an animal but me. They all had the deer in the headlight look in their eyes. This is why today I get calm in tight situations.

lee

On 7/31/2014 6:13 PM, Scott Bates wrote:

I have been on an estimated 150-200 fires in 48 years with a few years of not going on fires when the kids were small. I never kept track of how many fires I was on. I know that I have over two hundred parachute jumps because a log was kept on jumps by the jump base. A third of those were probably practice jumps. I've had some close calls but relish talking about them. If you were close to going into a fire shelter even if they didn't exist, you were in the wrong area of the fire for too long and someone probably screwed up. My job as a safety officer now is to reinforce the 10 Standard Fire Orders and 10/18 Watch Out Situations. We discuss the daily hazards and Situation Awareness at all times with having a lookout, good communication before engaging a fire. You have known and timed fire escape routes through the day and know where your safety zone is at all times. In short, LCES (acronym)is known about and communicated by and to all fire fighters through the day as conditions and locations change. We use trigger points of fire activity and location of the head of the fire and fire intensity to get fire fighters well out of the way and into a safety zone long before they can get themselves into trouble. We make sure that we have qualified leaders who have to carry a red card of fire qualifications before we let them be in leadership positions. Our safety system is a lot better than it used to be, but we still have fire fighters like the Granite Mountain shots at Yarnell that violated safety rules that they knew about and still get killed. Plus fire is inherently dangerous and you can't mitigate all circumstances.

Scott

On Thursday, July 31, 2014 5:01 PM, lee bates wrote:

Jay

On the Wheatland fire as the crew boss, I told the fire leader that I did not want to go into the canyon with my crew because it was a death trap. He said follow my orders or I am sending you home. My replacement probably would not have looked for a crown fire. From my experience of being on 40 fires by then, I knew it would crown if the wind came up. All 30 of the crew would have died because by the time I ran down from the canyon rim after the wind came up, the fire was crowning. As it was, we barely got to the vehicles and the radio to call for help before the fire caught up to us. You have never lived unless you had to run from a crown fire. We threw down all our tools. They are probably still there. One guy was just walking along. I said look behind you at the crowning fire. Only then did he take off running. I know I did the right thing by protecting my crew. I surveyed with and was the boss over 7 of the crew so I knew them personally. How would I have felt if I abandoned them as crew boss and they had died? I would have hated myself for the rest of my life. Ask

Dodge on the Mann Gulch Fire how he felt. I was the last of my crew to get to the top of the hill because I was counting heads. It got really hot. The fire leader was inexperienced. We were lucky that a B-17 was close had a full load and could plainly see our vehicles at the top of the ridge. I told my story at a Smokejumper Reunion in Missoula to the last surviving member of the Mann Gulch Fire, Robert Sallee, and he said I did the right thing. Also I told this story at the Dow's Smokejumper Reunion and they respected what I did. Scott was there both times. I love the B-17 because it saved my life. That is why I restored one at Boeing as a volunteer MRB Engineer.

Lee

On 7/31/2014 2:48 PM, Jay Bates wrote:

If I and my crew had been called up on that Wheatland Fire you went on, I would not have gotten my crew into that situation you found yourself in. We were not called because our location was too remote surveying in Hog Park a whole days drive back to Encampment.

From: lee bates

Sent: Thursday, July 31, 2014 9:48 AM

To: Jay E. Bates; lee bates; scott bates

Subject: Wild Fire Fighting mem 4 7-31-14

<u>Wild Fire Fighting</u>

I fought 40 wild fires when I worked from 1961 to 1963 for the Bureau of Land Management and Forest Service. My brother Scott has been on about 200 wild fires, my brother Jay 12 wild fires and my nephew Brett on many wild fires. A lot of my fire fighting was done on weekends so I earned a lot of overtime to use for my college education. My Memoirs tell of the Wheatland, Wyoming fire where I had to run for my life. As I write this in July 2014, there are 5 wild fires burning near Ellensburg where I live.

Dwight Lee Bates

The people around here who cleared the brush away for 50 feet around their homes had their homes saved. This was during the Taylor Bridge Fire in 2012. The 60 ones who did not clear the 50 feet lost heir homes. Lee

In the meantime, the fires are getting more intense because of the fuel build ups from lack of logging and lack of controlled burns to get the landscape back into a more natural state. There are more homes in the forest and on the edge of high intensity areas too. Scott

The first following photo shows Scott and Brett on a fire together. I got it from Scott. We kidded Scott that he had no dirt on him like Brett. The next two photos are from Brett's Facebook Page. Brett is the son of my brother Scott and is my nephew. He keeps an excellent Facebook Page that I follow.

Dwight Lee Bates

Dwight Lee Bates

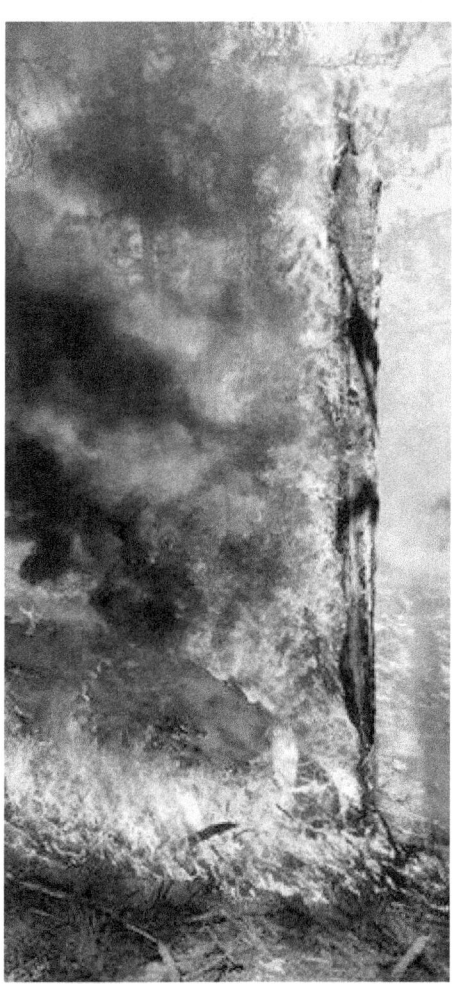

Dwight Lee Bates

California

Shasta

I have always admired Mount Shasta, shown in the following photo from Wikipedia, when I drove by the base of it or flew by it from the air. At 14,179 feet, it is slightly lower than Mount Rainier's 14,410 feet high. I heard that my brother Jay had climbed it. But he reported a bad experience. He just talked to an experienced climber on the mountain who soon fell to his death. This caused my brother not want to climb any more. I feel the same way when my guide on Mount Rainier was killed by an ice fall. He had told me a year before not to stop under the ice fall since it could come down any time. It did a year later killing 11 climbers roped up under it. Their bodies are still there covered up with tons of ice. Recently 6 climbers are missing and assumed dead on Mount Rainier due to an ice avalanche. After we climbed Mount Rainier, my Aunt Dorothy called me up after the 11 climbers were killed. She berated me for endangering the life of myself and my 2 brothers. She said they have wives and children and why did I not think of them before I evidently talked my brothers in climbing Mount Rainier? The slope of Shasta and Rainier are steep. You only have a few seconds to dig in your ice ax to prevent an ever increasing speed in your fall down the mountain. Since you build up a certain speed, you are not going to stop unless you hit a rock pile or fall in a crevasse. They asked the climber George Mallory whose body they have found on Everest why do you climb mountains? He said: " Because it is there." In other words, they are a challenge.

Sierras

All we hear about in Washington is about the Sierras in California. Many of our people climb there. Fred Beckey said at a presentation I recently went to that the Sierras was his most favorite place to climb. And he has many first ascents in the Cascade Mountains. My brother Jay climbed Mount Whitney with his son Jason. It is 14,505 feet which is 100 feet higher than Mount Rainier's 14,410 feet. But it is a walkup. I have photos showing Jay and Jason on the top of Mount Whitney. The shelter on the top is for shelter in the event of a sudden storm. The following photo, from Wikipedia, shows how rugged Mount Whitney is. It is the highest mountain in the lower United States. Colorado has 50 mountains which are

14ers or higher than 14,000 feet. Yet Mount Whitney is the tallest. Emily and Jason have gone backpacking in the Sierras. They have gone on many Bates' backpacking trips over the last 42 years. They grew up as they went with us. I remember when Jon and Brett caught their first fish. As a young boy, Jon asked me how you find your way around the wilderness? I said you navigate by landmarks and pointed some out. I have never been lost or even turned around. I look back to see what it looks like where I came from. This helps you find your way back. The next photo shows Yosemite Falls from Wikipedia. I went with Jay to Yosemite National Park one year after a business trip. I remember camping at the Hetch Hetchy River and unknowingly making a fishing pole out of Poison Oak. We slept up high on the rocks in case they let water out of the dam. The area was recently burned over by a forest fire. I think a camper started it. At Half Dome, you climb up the easy side by the use of cables. I think all of Jay's kids made it to the top. Half Dome is a main feature of Yosemite Park. The face of it is a technical climb. Jon got ½ way up El Capitan on the Yosemite wall in a snow storm. The rangers asked Jon if he wanted to be rescued as they were hanging off the wall on a portable ledge. They said no and they were having fun climbing.

Dwight Lee Bates

Dwight Lee Bates

Fred's and Dorothy's Cabins

When we visited Elk Mountain, my most favorite place was Uncle Fred's cabin. As a kid I loved going there. I inscribed my name on a Quakie tree which is still there. One night Aunt Dorothy and my mom spent a night there. They had cooked Mountain Sheep meat that night and threw the grease outside the cabin on the ground. During the night while they were sleeping, they were woken up by a screaming Mountain Lion on the roof of the cabin. The grease had attracted it. Then the Mountain Lion jumped off the roof and landed in some old bed springs. This caused him to fight it and let out more annoying screams. I remember when Fred built the cabin, he said he raised the roof beams by himself instead of using a front end loader. The cabin was not very big as it was 20 feet by 20 feet and 1 room. The logs were only about 6 inches in diameter so they were easy to handle by one man. I asked Uncle Fred how he built his cabin on the ranch. He said he went up into the timber with a draft horse. He cut up fallen logs with a hand saw. He then skidded the logs to a road. He borrowed Uncle Robert's truck and hauled the logs to the cabin site. Then by himself, he hoisted the logs up by lifting only one end. Later he had a young kid helping him. Aunt Dorothy had a cabin on a lot at the Bow Basin which she bought from Uncle Robert. It is about 10 by 20 feet as shown by the following photo. I only stayed there once. The site was on a pile of rocks and there were not many windows. I liked Fred's better. Dorothy also had a Bully Barn in which I slept during ma's funeral. It was even smaller at 5 feet by 10 feet. But it was nice inside. Uncle Donald had a cabin built by Fred's before he died. He used to escape there to get away from everybody. I guess my cabin is my 1967 Cougar Camper. At least I escape in it. The day it goes is the day I lose my freedom.

Dwight Lee Bates

Ingalls Lake

In 2011, I hiked to Ingalls Pass with Mike Harper and his son Justin. It was reviving a lot of the 12 trips we had done together. I had not seen Mike for about 5 years. We got a late start so the trail head parking lot was packed. All of Seattle loves this short but beautiful hike. Also it was on a Saturday which is the only time Mike could make it. It was the last day of the hiking season and hunting season was on. The campground at the end of the paved road was full. We saw a lot of hikers. The leaves were changing and it was beautiful. I hit a wall so I had to stop and eat an energy bar. Mike and Justin hiked to the small tarn shown in the picture. When we crested Ingalls Pass, the view of Mount Stuart was spectacular. I started back right away so I would not cramp up while Mike went on. They caught up with me later. I met them back at the trail head. We looked for a camping spot and found one close to the trail head along the creek. I had Kentucky Fried Chicken for everyone. As we ate by the fire, I told Mike I was going back to Boeing to work on the 787-9. It was October, so the sleeping was cold. We split up as Mike and Justin drove back to Seattle and I drove back to Ellensburg. I see why the hike is so popular. I have done it a number of times and it never gets old. I usually hike another 2 miles from the pass to Ingalls Lake.

The following is from a local trail guide.

"As with the Longs Pass Trail, you'll start out on the old miners road leading up the Teanaway Valley. Within the first 0.25 mile, the wide roadbed fades to a true trail. It also splits, the path to the left heading to Esmeralda Basin and Fortune Creek Pass. Go right to start your climb out of the Teanaway Valley. You'll be hiking through the first flora stratum you'll encounter on this hike, with huckleberry bushes, a few lilies, and other lesser known flowers, such as pipsissewa and wintergreens. At 2 miles, go left at another trail junction (right leads to Longs Pass). The trail angles upward, climbing steadily and at times steeply. As you near Ingalls Pass, the trail meanders through a meadowland. The final 0.3 mile switchbacks up to Ingalls Pass, about 3 miles from the trailhead. Here you'll enjoy spectacular views of Ingalls, and Esmeralda Peaks across the Teanaway River valley behind you." I hike to Ingalls Lake 2 miles from Ingalls Pass shown in the following pictures.

Dwight Lee Bates

Dwight Lee Bates

Dwight Lee Bates

Tamarack

I like to hike in October in the mountains and go to Elk Camp because of Tamarack. These beautiful needles are shown in the photo from my hike to Ingalls Pass. The needles fall off and leave the tree bare. I notice Tamaracks or Larches when I go to Elk Camp each year as explained in a story in these memoirs. The needles would fall off the trees into my Cougar car cowling when I was at Elk Camp. I thought that one of the reasons I moved to Ellensburg to retire is my love of the Tamarack. The tree is really prevalent in this area. You seem to have to drive to high altitudes though to see them. Their natural range is in Canada. I like Quakies also which are in a grove that we own by our house in Ellensburg. These two trees to me make up the beauty of the West. I always associate hunting with Tamarack. I use the time when the needles fall off to gage time. The only problem is when the needles get in the way. The Evergreen Trees do not shed their needles like Tamarack does. The bare branch adds to the ruggedness of the tree.

Dwight Lee Bates

Lake Melakwa

In 1968, Diane and I hiked to Lake Melakwa near Snoqualmie Pass in Washington. I carried a day pack and Diane carried nothing. As I drive by on the freeway, I always look over to see the high peaks above Lake Melakwa. The hike is popular since it is only 4 miles to the lake with an elevation gain of only 2,180 feet. This makes it an easy day hike. You can eat lunch at the lake. On our hike, a guy came down saying he had seen a Black Bear. I had to talk to Diane to keep her going up the trail. Her usual banter is: " I do not want to be beaten up by a bear.'" She hates bears and bugs. We kept on going past the beautiful Keekwulee Falls. This sounds like an Indian name. It probably means hike for stupid hikers. The hike was steep in places so we took our time while all the time looking out for the bear. When we got to the lake, we had it to ourselves. We ate lunch and admired the big granite peaks above us. These peaks are climbed quite a bit since they are readily accessible from the highway. Chair Peak is my favorite because you can see that it looks like a chair you can sit in. However it is a dangerous short climb. People get in trouble here all the time per the news reports. The Tooth is another favorite climb that looks like an incisor tooth. Then the closest peak is Denny Mountain. You ski off the top of this peak at the Alpental Ski Area. Diane and I took ski lessons there. No we did not ski off the top of Denny Mountain since it is a double diamond expert run. My nephew Jeff has skied it. A woman was killed here when she skied under the cable barrier and directly off the top. I am sure that she did not want to do that. Also the mountain gets a lot of small plane crashes. A friend of mine who lives there said if he hears an airplane engine, he listens for a crash. If he hears one, he calls the sheriff. You have to turn at Snoqualmie Pass. It looks like you should fly straight ahead but it is a trap. The locals know about it but new pilots to the area may not know about it. I always fly high over the pass and not through it. It is a lot safer but not as scenic this way. I would recommend the Melakwa Lake Hike for beginners and old men like me now.

Melakwa Lake

From Wikipedia, the free encyclopedia

Coordinates: 47°27'1.16"N 121°28'8.27"W

Melakwa Lake is a lake in King County, Washington. The name Melakwa comes from a Chinook term for "mosquito".[1] It is located along the Pratt River just below the river's true source.

Contents [hide]
1 Access
2 Characteristics
3 See also
4 References
5 External links

Upper Melakwa Lake

bordered by Chair Peak (to the right) and Kaleetan Peak (to the left)

Location	King County, Washington, USA
Coordinates	47°27'1.16"N 121°28'8.27"W
Primary inflows	Pratt River
Primary outflows	Pratt River
Basin countries	United States
Surface elevation	4,505 ft (1,373 m)

Access [edit]

It is off of I-90 and can be accessed from the Denny Creek trail head. It is accessible via a moderate, all-day hike from Denny Creek to Melakwa Lake.[2]

Characteristics [edit]

The lake is located in a narrow valley, with Chair Peak to the east and Kaleetan Peak to the west, and is usually accessible by late June. It is known for its clear water and beautiful blue-green color. A short scramble up the talus slope on the north end of the lake leads to Melakwa Pass, where views of Gem Lake, Glacier Peak, and the North Cascades are visible on clear days.

ROCKHOUNDING

Rockhounding

I have gone on rockhounding trips with my brother Jay's rockhounding club in San Francisco many times. I publish my brother's following trip report to show what a typical trip consists of. This trip was for Turquoise at Tonopah, Nevada. My brother Jay led this trip. I wrote a story about a trip Jay and I took to Southern Oregon rockhounding called:" 5 Days In Southern Oregon." Jay had it published at his rockhounding club and it was well liked. It led to the club coming to Oregon to rockhound. The photo shows my brother Jay which I took off the TV when he was on the national TV Show "Cash and Treasures." He was so excited when I called him after the TV program aired, that he could hardly talk.

Subject: Re: Tonopah trip Report for Mineralog. (Credit for article is by Jay Bates and Mary White)

Deanna and Jay,

Jay should be the first author on this, as it starts with his description of arriving at the campsite (plus it is mostly his work!). Nice job, Jay.

Mary

Turquoise and Monte Cristo Trip near Tonopah, By Mary White, Jay Bates

Sometimes you just need to leave the finer things behind and go back in time to simpler surroundings as you remembered in your youth. Often it involves meeting the natural elements on their own terms and reconciling your own existence within the vastness of the universe. Such primordial urgings can and often do, goad us to go into the wilds and search for nature's precious and wondrous gemstones. One such adventure started out on Sunday May 25 by a memorable drive through Yosemite National Park with the Dogwoods and Redbuds in Spring bloom wafting heavenly fragrances, among exfoliating granite monoliths and sparkling tarns. Leaving Yosemite we traveled down the eastern flanks of the Sierras and out onto the thirsty breaks near Mono Lake, among the dusty Ponderosa pines and the omnipotent presence of fourteen thousand foot peaks of the eastern Sierras and Boundary Peak in Nevada. Your spirits soar among such magnificent surroundings.

Dwight Lee Bates

Picture Agate

Picture Agate is a rock that has a scene on it. The following photo shows a picture agate that I collected from Weiser, Idaho. It shows the Amazon River running through a jungle. The river has a sandy beach. The jungle is across the horizon. The sky is in the back ground. I had a belt buckle that showed a boat out on the water. People really liked that specimen and wanted me to give it to them. It was a favorite when I showed it during rock shows. My brother Jay has a lot of Picture Agate in his collection. In fact a week ago, we were collecting it above Succor Creek in Oregon. We went there to rockhound and fish for Brown Trout. This story called Owyhee River is written up in these memoirs. Jay and I collected Wonderstone in Nevada. These specimens had lines running through the material. I made a belt buckle out of Bruneau Jasper. It has circles in it and takes a good polish. It came from a commercial mine in Idaho. Specimens that people like have landscapes on them. They usually show mountains and the sky. It is rocks like this that make you love rockhounding and make you want to come back for more. I collected one specimen that if you looked at it with a magnifying glass you could see that the images were made up of smaller images.

Dwight Lee Bates

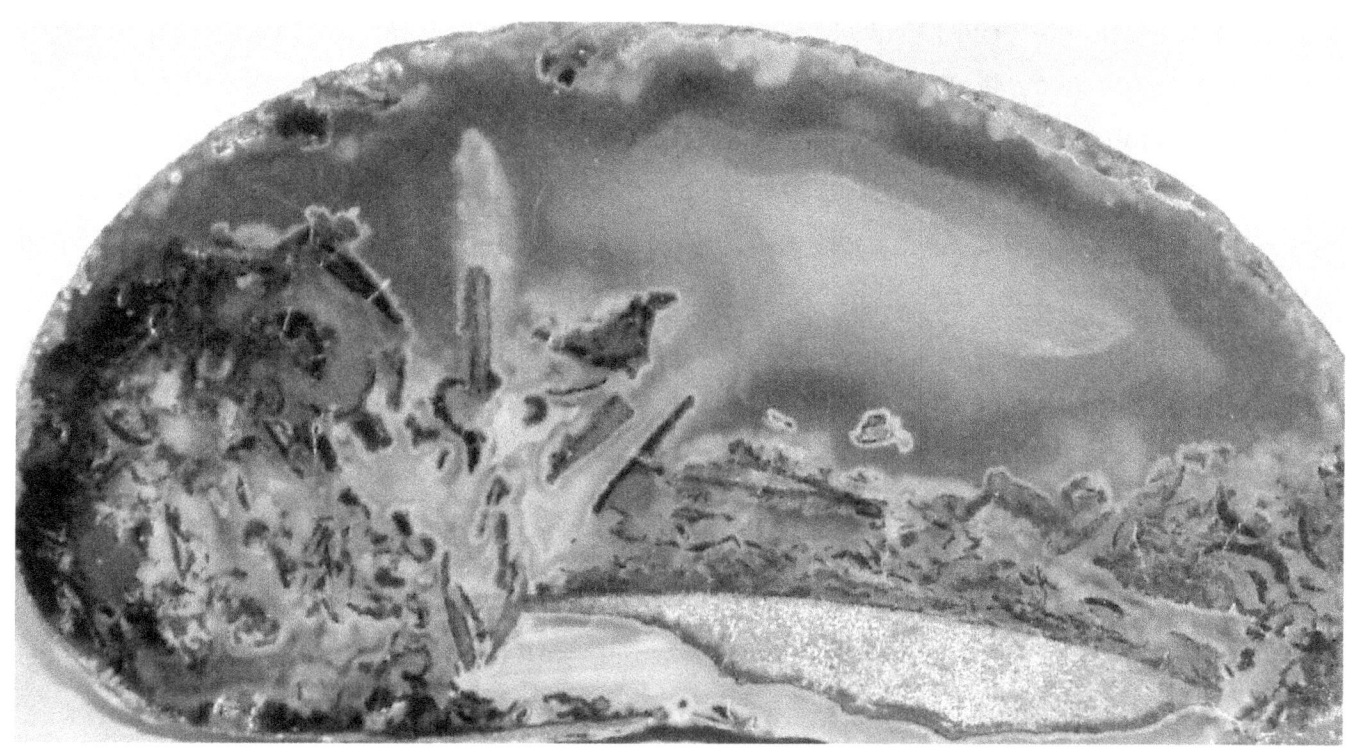

Hampton Butte

In 1998, Jay and I left our Maury Mountain, Oregon camp site and drove to Hampton Butte about 20 miles away. It was wide open Pinion Pine area known for its green Petrified Wood. It is in the form of logs or limbs. As shown in the photo, it is a great free camping area. It even has an outhouse. You have to dig deep for the wood. There are numerous pits already dug. Later when Jay's rockhounding club from San Francisco came here, we found 3 Jolly Green Giants that I called them. They are the beautiful Green Petrified Wood shown in the following photo I took. I found a piece of this about the size of your fist on this first trip. I walked around looking for float but found a horse trailer instead hidden away in the woods. I think it was stolen and someone hid it there until the heat dies down. A friend of mine said you could find rounds of Petrified Wood on the surface but I did not find any. A member of Jay's club asked me how I could navigate in the woods. I said the sun is in the west. I traveled east so I came back toward the sun or to the west. I hit the fence first which I knew crossed the road so I followed it to the road and back to camp. We sat around the campfire with Jay's club. We talked about Area 51 the U S Government secret test area which a member had gone to. Someone said we need tin foil hats which we made as a joke. Most people dig here in the morning before it gets hot. We never found a complete log like we did at Congleton Gulch near here. I had trouble with the Cougar here when a rubber gas hose you could not see collapsed in on itself. I had to go into Prineville to get another hose. I gave a choice piece of wood to Harry of Jay's club for jewelry and have been his good friend since.

Dwight Lee Bates

NEAT PLACES

Dwight Lee Bates

Elk Mountain

My relatives on my mother's side were born and raised in Elk Mountain, Wyoming. It was a nice place to visit when we lived on the farm since it was about a 4 hour drive. My Uncle Fred had a cabin there I liked to visit. My name is still carved on a Quakie tree by his cabin. My mother and brother Scott lived there in Aunt Dorothy's house for a while. It is a small town with a trading post for the only store in town. It has an old lodge and church in which we had the service when my mother died. I remember going to Elk Mountain to the dances when I worked on the A-1 Ranch during the summer in 1959. We danced with the available girls which were not many. We did the Bunny Hop Dance which made the floor bounce. The dance hall was called the Garden Spot Pavilion. My mother said all the great bands came to Elk Mountain to play. I did not believe it until I did research on it and found out it was true. I think the famous people just wanted to see what country folk were really about. I asked my Uncle Fred about the big bands like Lawrence Welk and Paul Whiteman playing at Elk Mountain. He said it was on the way to Denver and the entertainers liked the small town feel. He said he and his college girl friend won the best dance partners at a Lawrence Welk dance at the Garden Spot. As a reward, Fred got to lead the Lawrence Welk Band as a conductor. The band as a joke played really fast on purpose, so poor Fred was waving the baton as fast as he could. Uncle Fred was glad that I was writing these memoirs on Elk Mountain. I remember driving back to the ranch after dances at the Garden Spot when the driver which was not me had too many beers. It seemed like we slid around every corner. I remember Bill Richardson's Piper Cub Airplane tied down next to the road. I would have loved to fly in it. At the dances, there were fights every Saturday night which made for good entertainment. The bad thing was water was gotten from the town well. You had to hand pump it out of the ground. Also there was only an outhouse with a Sears and Roebuck catalog for wiping. But I used to read it while I was sitting there. Now that I used to live in a big city, I love small towns like Elk Mountain. That is what made me retire to Ellensburg which is a small town like Laramie where I went to college. The following photo from Wikipedia shows the Elk Mountain area.

Wikipedia and: www.elkmountainhotel.com/history.html and: www.wyohistory.org/encyclopedia/elk-mountain-hotel-and-garden-spot and http://www.over-land.com/ and http://www.wyomingtalesandtrails.com/stagelines.html and Nancy Anderson of the Hanna Basin Museum www.hannabasinmuseum.com andvictora2@carbonpower.net are the sources for the following:

Elk Mountain is a town in Carbon County, Wyoming, United States. The population was 191 at the 2010 census. The town shares its name with a mountain 7 miles northwest of town. The hidden gem of Elk Mountain was named after a Sioux chieftain called Standing Elk. The Town is hidden in a small, tree covered valley, just 3 miles south of exit 255 of the Interstate 80. Elk Mountain is located in Carbon County in SE Wyoming almost centrally located by Laramie, Rawlins and Saratoga. It is 90 miles NW of Cheyenne, WY and 154 miles NW of Denver, Colorado. The town is perched at 7,264 feet and lies in the shadow of its 11,156-foot mountain. The settlement of Elk Mountain is a quiet community of cottonwood tree lined streets, and picturesque buildings, with world class trout fishing right downtown. Entering Elk Mountain is like stepping back to a more leisurely, less pressured time. You find a hamlet nestling amongst a forest of Cottonwood trees that must be almost unique in Wyoming. Much of the historical significance of the Elk Mountain area lies in the development of the transportation network linking the east and west coasts.

The Medicine Bow River crossing was used by the John C. Fremont expedition of 1843. On August 2nd of that year, Fremont's party camped in the proximity of the "Medicine Butte", an early name for Elk Mountain. The river would become a major crossing for emigrant and stage-coach travelers.

The Stansbury expedition in the 1850, led by famed mountain man Jim Bridger, crossed the Medicine Bow farther north seeking a route for wagon travel. Later in1856, Lt. F.T. Bryan, realized the potential of Stansbury's route and suggested it be used for the Overland Stage started by stage coach king Ben Hollady. By 1862, the operation was imperiled by constant Indian attacks. Holladay chose to move the line southward, back to the Medicine Bow River Crossing, where he built a stage stop. In 1862, Fort Halleck was built on the Overland Stage route a few miles west of Elk Mountain to protect travelers passing through this region. The fort was named after Major-General Henry G. Halleck, a key military aide to President Lincoln. The government maintained the fort from 1862 to 1866 when it was decommissioned because the Indian threat had diminished. The owner of the stage stop found a sufficient volume of trail traffic to maintain a toll bridge, although eventually stage traffic waned.

During the years spanning 1862 to 1868 20,000 emigrants a year moved west along the Overland Trail. Elk Mountain's first mercantile store was constructed in 1902 using lumber from the Carbon Timber Company.

In 1905, the Elk Mountain Hotel was built by John S. Evans on the property previously used by the Overland Stage Station. The building's architecture is Folk Victorian style reminiscent of what was found on the frontier during that time. Adjacent to the Hotel stood the Garden Spot Pavilion. The Garden Spot was host to such famous entertainers as Louis Armstrong, Tommy Dorsey, Gene Krupa and Lawrence Welk. These entertainers inspired hundreds to "jump on and ride" the Garden Spot's magical dance floor during the 1950's. Both the Hotel and the Pavilion are listed on the National Register of Historic Places but sadly the Garden Spot Pavilion was closed. It was deemed un-restorable and was demolished for safety reasons.

The Hotel property served as an important component in the economic and social life of the Elk Mountain community. The lodging, mining and livestock industries boomed. The property enjoyed a steady clientele and became a way-station for entrepreneurs and laborers who traveled here for the timber, mineral and ranching industries.

Today the comfortable and welcoming Inn still stands where it was first constructed over100 years ago and welcomes guests' year-around.

Sitting as it does along the banks of the Medicine Bow River, Elk Mountain is a powerful draw for trout fisherman from across the world.

Visitors interested in beautiful scenery should take the round-the-mountain drive on Pass Creek Road as it is an area of stunning vistas teeming with wildlife and provides a real sense of adventure without too much inconvenience.

Dwight Lee Bates

Fort Halleck

My Aunt Dorothy told me about the history of Fort Halleck. She learned about it when she did her University of Wyoming Masters Thesis. I have always been interested in the fort. One day I would like to visit it even though it is on private land.

Wikipedia:

Fort Halleck was a military outpost that existed in the 1860s along the Overland Trail and stage route. It was in what was then the Territory of Idaho, now the U.S state of Wyoming. The fort was established in 1862 to protect emigrant travelers and stages transporting mail between Kansas and Salt Lake City, Utah. It was named for Major General Henry Wager Halleck, commander of the Department of the Missouri. He was later General-in-chief of the Union armies. The fort was located on the northeast side of Elk Mountain at an altitude of about 7,800 feet. At the time, the area around the fort was well watered and well stocked with game. The fort was reasonably large with stables for 200 horses, company quarters, office quarters, and a hospital and surgeons on staff. The Overland Trail was established in 1860 following the same general path as the Cherokee Trail. This trail was in use in the late 1840s by miners heading to California. In 1861, the government moved the official mail route to the Overland Trail from the Oregon Trail. This was due to threat of Indian attack. The mail contract was assigned to Ben Holladay who established a stage line and stage stations along the Overland Trail. Pressure by white immigrants and shifting buffalo herds forced the Indian tribes to the Laramie Plains. Here they came into conflict with travelers on the Overland. Soldiers from the Eleventh Ohio Volunteer Cavalry stationed in Camp Collins in Colorado were dispatched north to build Fort Halleck. This was to protect the trail from Camp Collins to the Green River stage station in the west. The trail was at its busiest in 1864 and 1865. During this time troops were often used as escorts and drivers for the stages. At times in 1864 and 1865, ongoing attacks caused the mail to accumulate at stations in Colorado and at Fort Halleck. Then it was transported to Green River via government wagons. The fort was abandoned in 1866 after Fort Sanders was built near the present day city of Laramie. Three years later the Transcontinental railroad was completed (mostly along the same route as the Overland Trail. It ran from Laramie to Salt Lake City). As a result, travel on the trail declined to nearly nothing. The outlaw L.H. Musgrove was charged in 1863 with murder at Fort Halleck. He was taken to Denver, Colorado, where he was lynched in 1868 by a vigilante committee. Today the fort site is located on private land. Only a single building (thought to be the blacksmith shop) remains standing. A marker is located on the cemetery.

Fort Halleck, Wyoming, was established in 1862 to protect the Overland Trail Stage Line from continuing Indian problems. The site chosen in a gap on the north side of Elk Mountain at an elevation of about 7300 feet, was near a spring with plenty of wood for cooking and heating. The fort was located in the midst of some of the most beautiful tall grass meadows along the trail, abounding with large herds of elk, deer and antelope. The fort complex was quite substantial, consisting of stables large enough to hold 200 horses, storehouses, two sets of company quarters, officers' quarters, a store, bake house, jail and hospital.

Exact accounts vary, but it seems as though a wagon train moving through the area in mid-summer 1862 was well supplied with "frontier" whiskey. The stationmaster proceeded to sell canteens full of the whiskey to the soldiers for $5.00. It wasn't long until many of the soldiers, including the entire night guard, were totally drunk. The commanding officer, Major O'Farrell, gave orders to search every wagon in the train, find the whiskey, and destroy it. The barrel was found, and the remaining contents were spilled out onto the ground. Unfortunately, this spot was right above the spring, and the whiskey went directly into the water supply for the fort. The soldiers wasted no time in saving all the whiskey they could, using whatever cup, canteen or camp kettle they could find. Some just lay and the ground and lapped it up! The gap in the mountains was called after that incident Whiskey Gap--a name which remains to this day.

The post surgeon at the time, Dr. J. H. Finfrock, kept very detailed records of the emigrants passing through the Fort Halleck station. In 1864 he recorded that there were over 4200 emigrant wagons, with a staggering number of 17,584 emigrants and an even more astonishing total of over 50,000 animals traveling the Overland Trail.

A diarist, Franklin E. Adams, who kept quite a thorough diary of his trip on the Overland

Trail in 1865, notes that soldiers from the Ohio Volunteers were stationed at Fort Halleck, and were paid $16.00 per month to fight the Indians. This stretch of the trail, from Fort Halleck to Sulpher Springs in the west, was considered to be one of the most dangerous, with regards to the Indian attacks.

On one occasion, Jack Slade, in charge of the Mountain Division of the Overland Trail and stationmaster at Virginia Dale, is reported to have gone into the fort store and used the canned goods on the shelf for target practice. After a second time of creating general mayhem at the fort, the commanding officer arrested him and refused to release him until Ben Holladay promised that he would be fired from the stage company. Jack Slade was fired soon after this incident.

Fort Halleck was abandoned in 1866, just four years after being established, and by the next year diarists described it as "the most dreary place on the entire route." One building remains in the area that some feel may possibly be the blacksmith shop. A stone marker erected by the DAR in 1914 marks the site of the Fort Halleck cemetery.

Bloody Lake Massacre

In about 1956, Aunt Dot and Aunt Lena took us to Bloody Lake, Wyoming near Hanna to look for arrowheads. Jay found a broken one. My Aunts told us a story that was not true about the massacre that took place at Bloody Lake. They told us that the wagon train guide feared an Indian attack and went up to the top of hill to sleep and witnessed the attack. All the settlers were driven into the water and killed. It was a freight wagon and not a wagon train like they told us (see the story below of the pioneer woman Mary Jane Bloomer Morey Stimpson Richardson for the true story. This shows how history as told from mouth to mouth and generation to generation is inadvertently changed). Also The photo of Bob Johnson by the pioneer's grave is not on the ridge above Bloody Lake. It is on Turtle Rock Ridge about 10 miles away. It looks like the pioneer woman wanting to be buried above Bloody Lake did not happen and is a wrong story. Bob said that she wanted to be buried on Turtle Rock Ridge. I tried to find records of this massacre at Bloody lake. But like Jay says it got lost in time. These errors point out that you need to do due diligence when researching for writing a book. You need to verify everything.

This is the true story on the massacre:

In October 1868, my mother Alice Willing Bloomer was on her way Fort Halleck from Percy Station to pick up some supplies when she witnessed a band of Indians attack a freight wagon train of five men hauling ties to Percy Station. The Indians killed three of the men and two others managed to shoot their way into a ravine and escape towards Fort Halleck. The Indians drove the oxen into the lake and hamstrung them (cut the backs of their legs) and left them to bleed to death in the lake. They next took the mules and fled before reinforcements arrived from Fort Halleck. This was known as the Bloody Lake Massacre. The three killed were the first burials at the town of Carbon. Her great great grandmother saved arrows from the event and they passed down through an uncle's family. Alice died four years later on 01 August, 1872 and was buried in a shallow grave surrounded by sandstone slabs stacked on top of each other on "Turtle Rock" Ridge (Note: we are currently working with the BLM to erect a historical marker at her grave site). The Alice Willing Bloomer partial story below was obtained from the grave stone in the photo,

Wikipedia and Ox Cart trip - RootsWeb and .genealogytrails.com/Wyo/carbon/bio3.html and Dwain Romsa.

Mary Jane Bloomer Morey Stimpson Richardson

My great great grandmother was named Mary Jane Bloomer, born in Mina, Chautauqua, New York in 1842, she was married three times. Her first husband, James Morey, was a Civil War officer serving under US Grant. She moved to Galena, Illinois prior to the Civil War where one of her Aunts lived and became a friend of the Grants and served for a time as Grant's personal secretary during the Civil War. Her husband was shot in the head, but survived. His discharge papers stated "Brain Hernia". He later died in an Insane Asylum in Washington DC and is buried at Arlington.

After the Civil War in 1867, Mary Jane (abandoned her insane husband) and her mother Alice Willing-Bloomer (also widowed during the Civil War) WALKED from New York to Fort Laramie with an oxcart and Mary Jane's three year old son Frank Morey, born in Ohio during the Civil War.

The ladies knew General Grant, General Rawlins, General Sheridan, etc. from the Civil War era and whom also happened to be in charge of the Transcontinental Railroad Construction. (We assume they were offered work before leaving for Wyoming on such an epic journey into the "Wild West". Can you begin to imagine the fortitude it took for two ladies with a child to do this in 1867 traveling alone in the wildness with only their wits for protection?). In Oct 1868, mother Alice Willing Bloomer was on her way to Fort Halleck from Percy Station to pick up some supplies when she witnessed a band of Indians attack a freight wagon train of five men hauling ties to Percy Station. The Indians killed three of the men and two others managed to shoot their way into a ravine and escape towards Fort Halleck. The Indians drove the oxen into the lake and hamstrung them (cut the backs of their legs) and left them to bleed to death in the lake. They next took the mules and fled before reinforcements arrived from Fort Halleck. This was known as the Bloody Lake Massacre. The three killed were the first burials at the town of Carbon. My great great grandmother saved arrows from the event and they passed down through an uncle's family. Alice died four years later on 01 August, 1872 and was buried in a shallow grave surrounded by sandstone slabs stacked on top of each other on "Turtle Rock" Ridge (Note: we are currently working with the BLM to erect a historical marker at her grave site). After returning to Elk Mountain, we have numerous more stories about this amazing lady including fighting off three renegade Indians and killing one in hand to hand combat at the ranch, "engineering" the irrigation for the Medicine Bow River Valley, building a post office, challenging a neighbor to a gunfight when he threatened the family, and several more. Submitted by researcher Dwain Romsa.

Dwight Lee Bates

<u>The Smokejumper Reunion</u>

 Ten times I have gone to the Smokejumper Reunion at the Dow's house near Ellensburg. It is located one half way up a ridge near Ellensburg Pass in Washington. About 30 ex Smokejumpers show up each year to gather fire wood for the Winter, and build buildings for the Dows. In turn, we are rewarded with tall tales around the dinner table and camp fire. Also we eat like kings. I always bring two Perkins pies which is appreciated by Lisa Dow. I put siding on a storage shed for 4 years, built a railing, built a step, stacked firewood and built a wall for the dining building. Doing this, I learned carpentry from Smokejumpers who do it for a living. My brother Scott who used to Smokejump and I sleep at the bottom of the ridge in a tent and my Cougar Camper. One year, I read a poem I wrote about the reunion which was well received. Last year, I told the story after dinner about the close call I had at the Wheatland, Wyoming forest fire. I like visiting with Murry Taylor who wrote the best selling book on Smokejumpers called: "Jumping Fire." Don Bell always tells good stories every year. I wrote a poem about Murry for which he humbly commented: "I never thought that anyone would ever write a poem about me." I always bring rocks for Julian the Dow's boy every year. Lisa Dow some times wears the gem stones I bring for jewelry. One year I gave her a beautiful red Sunstone I found in the desert near Plush, Oregon to wear around her neck. Just before everyone goes home, we play a betting game where everyone puts 20 dollars in the pot which we flip for. Then we all line up for a group picture. It is one of my favorite events for the year. My brother Scott and I flew over it one year in my Cessna 172 and other times I like to fly over the beautiful area.

Elk Camp

Every Elk Hunting season at the end October, I go to our Elk Camp. It is located at 5,000 feet up a canyon on the North East edge of the Kittitas Valley. I always bring two Perkins Apple pies for desert. I camp in the Cougar Camper at night which is always cold. We eat like kings and I always do the dishes. I do not hunt but like the camping experience and the camaraderie with the hunters. One year I laid down in the cook tent as the hunters were hunting the first day. I heard the sound of running Elk through the ground near the camp. I ran outside to see a herd of Elk run by. When the hunters came back for lunch, I asked if they had seen the Elk herd. No one believed me until I showed them the hoof prints in the soft dirt. We build a fire and sit around it telling tall tales. We have our own shower and out house. We have two tents - a living tent and a cook tent. Chopping firewood is also one of my chores. One year the hunters arrived with the back door of their trailer wide open. They found that a $5,000 rifle had fallen out. I got the owner and we drove down the road but someone had found it. We went to the Sheriff to report it lost. Later someone returned it because when they test fired it, they could not hit any thing with it. I used to tell my Sasquatch story just before we went to bed to keep the hunters awake. One year someone shot an Elk near camp on the road but usually we come up empty. I do not get any meat since Diane doesn't like Elk meat.

Owner Bob Schnebly's Comments:

I started the Elk Camp in 1997 for hunting Elk as a way to spend 2 weeks with my son just to relax. I chose the spot because it is a wide flat spot, near a creek with water and the beautiful Tamarack or Larch Trees. I enjoy the fun with my friends around the campfire and dinner table. At night, I love how close you are to the stars. It is about 10 miles into it. Usually 8 people hunt Elk and 20 people visit - some of which stay for dinner. We eat good. If you lose weight in Elk Camp, it is because you are hunting too hard. We have 2 tents of canvas and steel framework. We used to bring up 2 wood stoves for each tent but have joined the 2 tents together. By doing this, we can heat both tents with a fan between them and a single stove. Elk Camps are used all across the country. It is a unwritten code that you can use the same spot every year. We usually get an Elk every 10 years.

Dwight Lee Bates

Great Falls

We lived in Great Falls from the time I was 3 in 1946 to 8 years old in 1951. We went back to Great Falls in 1998. I went directly to our third house due to my excellent memory. Then we went to Emerson Grade School. Here I went to the First, Second and Third Grade. We moved to the farm at Ault, Colorado during my Third Grade as I described in these memoirs. I went inside the Emerson Grade School and found that it had been converted to a Head Start Office. But I found the room I had my grades in. I even found the hook that I used to hang my coat up on. I can member the coloring books and the tests where you had to match similar objects. I had to put on those stupid overshoes that went over your regular shoes. These would never go on or come off easy. I remember being the first to recite the alphabet in the first grade. Then we had to learn the multiplication tables and I was first in my class again. But I used to practice them at home. Don't tell anyone because they thought I was smart. It was an Engineer in the making. I loved The Weekly Reader our own newspaper for kids. I can remember recess. There was a Diary Queen across the street. We were told we could not eat on the play field during recess. So we leaned over the fence as we ate our ice cream cones from the Dairy Queen. The winters were tough. We had to walk about a mile to school. Of course you had to wander around making the stops. There was a nice old lady who would give you lemonade and cookies if you charmed her. Then there was the candy store. Five cents went a long way then. I can smell the smoke from the burning leaf piles now. I loved school. I remember the fire drills. We played dodge ball during recess. You would stand next to the brick wall and they would throw basketballs at you. If you were hit, the person who threw the ball would stand in front of the brick wall. I would flirt with the girls. They would giggle and say something to each other that you could not hear. I always imagined it was:" What a handsome little fellow he is." We rode our bicycles 2 miles downtown all the time. We thought nothing of it. Jay had a Schwinn Bike which had a horn on it. He stripped it down to save weight. Intact they are worth a lot of money these days. My el cheapo bike was always breaking down. Now a days the kids are driven every where. Every Saturday, we would ride our bikes 2 miles to the movie theater downtown and watch the serial movies. The hero wearing a white hat would fall off the cliff while chasing the bad guy Black Bart and the movie would stop. Then a message would appear across the screen. Kids return next time to see what happens to our hero. Gee and it was just getting good. What do you expect for 10 cents? The next week the hero grabbed a bush on his way down the cliff face and was saved. Imagine that. I thought my hero was dead for a whole week. These kids now a days with their video games do not know what they are missing! We were in the Cub Scouts. Jay needed a project for his Wolf Badge. I said no problem. I whipped out a wooden ship from an old apple box. I used tree branches for smoke stacks. It was the Queen Mary. Then I saw it in the window of a store with a ribbon on it. So at the supper table I said: " Since I built the ship, do I get to have the ribbon?" I heard: " Shut up and drink your milk." The story of my life. When I want something, someone says shut up and drink your milk. Ma came to my Second Grade Class as a visiting mother. She said: " Now kids turn around and make faces at the kids behind you." I used to do this on my own so I did not see what was so educational about it. I remember making puppets for Cub Scouts. We carved the heads from balsa wood. We put a hole in the center of the head for your finger to go in. Ma sewed up dresses for the puppets to hide your hand. They must have been female puppets but looking at mine you could not tell. The stage was made out of a cardboard box. God the puppets were ugly! I carved mine from the Frankenstein movie I saw I think. I guess the project was for another one of Jay's Cub Scout badges. They should have taken away his Wolf Badge for those ugly puppets. My brother Jay had a friend next door called Dennis. They got in a fight. That is another story. I remember going to the "Great Falls Select" Baseball games. They were named after a local beer brewed in Great Falls. I went to try out for a Little League Baseball team. The big boys threw me the baseball as hard as they could. It stung my hands after I caught it. So I told the big boys that. They said I was chicken. I was bullied at an early age. Maybe this is why I turned out so well. I liked following dad around as he was golfing at the Meadow Lark Country Club. I remember the plastic tees I would pickup. We wandered over and around a little creek. The Meadow Larks were always singing. And when I found a ball dad would say: "Good boy" when I gave it to him. Maybe this is why I love golf today as written up in these memoirs. It is such a beautiful place. Who cares if that little white thing does not go where you want it to. I continue to look for golf balls today either because my dad taught me to be cheap or he is looking down from heaven and saying: "Good boy." In

Cheyenne, he made us dive down in the golf course lake after hours to retrieve golf balls. We got 2 big cans of them that no one ever used. Maybe we got them because they were free. I could imagine the President of the Club like in the movie " Caddy Shack" catching us. He would say to dad : "You know we sell new balls in the golf shop." I remember the first snowmobiles. Actually they called them snow machines. They had an enclosed cab and an airplane engine with a prop behind the cab. You steered them with the ski in front. When the streets got covered with snow, you would see these machines going up and down the streets. I think Great Falls was a great place to grow up. It must have been since I still have fond memories of it.

Northwest Trek

In 2012 and 2013, when I was working for Boeing Frederickson building parts for Boeing jetliners, I used to go to Northwest Trek. It was a way to relieve the stress. I got the Senior rate so it did not cost very much. Also I rode the zip line there as described in these memoirs. A photo I took shows a bobcat whose eyes glowed yellow when I photographed him. You walk along wooded paths looking at the animals in their natural state. They do not look like they are in cages. The first photo shows a Wolf looking content and relaxed. They never looked this way when I saw them in the wilds. I liked to hear them howl when I flew into high mountain lakes in British Columbia by float plane. I took another photo of Bison near the lake. I framed the shot in a fence with Elk horns. We saw a Mountain Goat which was right next to the road where our rubber tired tram went by. I have seen them in Glacier Park but not so close. I took my brothers to Northwest trek after we climbed Mount Rainier in 1980. Also I liked to walk the trails to get under the zip lines shown in the following photo and watch people slide down the wire. It was only a 30 minute drive from my apartment so I went to Northwest trek as much as I could. I am glad someone has preserved the Pacific Northwest animals in their natural state.

Dwight Lee Bates

Dwight Lee Bates

THE END

Please do not regard this book as the end. I have already written 2,000 more pages for my Memoirs in the same format as this book. They are arranged in book form now and I can easily move the stories around. If you like this book, you will like my other 5 books in the series of my memoirs. It all depends on how this book sells but the other books can easily be released to the publisher. Also I plan to publish this book in e book format for the use of Kindle to see how it sells in that format.

Dwight Lee Bates

Dwight Lee Bates

Dwight Lee Bates

Dwight Lee Bates

Dwight Lee Bates

Dwight Lee Bates

Dwight Lee Bates

Dwight Lee Bates

Dwight Lee Bates

Dwight Lee Bates

Dwight Lee Bates

Dwight Lee Bates

Dwight Lee Bates

www.ingramcontent.com/pod-product-compliance
Lightning Source LLC
Chambersburg PA
CBHW080233180526
45167CB00006B/2263